THE ULTIMATE GUIDE TO

PERMACULTURE

THE ULTIMATE GUIDE TO
PERMACULTURE

WRITTEN AND
ILLUSTRATED BY
NICOLE FAIRES

Skyhorse Publishing

Skyhorse Publishing books may be purchased in bulk
at special discounts for sales promotion, corporate
gifts, fund-raising, or educational purposes. Special
editions can also be created to specifications. For
details, contact the Special Sales Department,
Skyhorse Publishing, 307 West 36th Street, 11th Floor,
New York, NY 10018 or info@skyhorsepublishing.com.

Skyhorse® and Skyhorse Publishing® are registed
trademarks of Skyhorse Publishing, Inc.® , a Delaware
corporation.

www.skyhorsepublishing.com

10 9 8 7 6 5 4 3 2 1

Library of Congress Cataloging-in-Publication Data is
available on file.

ISBN: 978-1-61608-644-2

Printed in China

Dedicated to my husband John,
who still doesn't let me quit,
and my three crazy girls. I hope someday
they look back
at our bus years with love.

Special thanks to:
Renato Faustini, for the amazing solar house models;
Skyhorse Publishing, for making all my dreams come true;
my editor, Jenn McCartney, for being awesome.

Contents

Introduction

The fish trap exists because of the fish. Once you've gotten the fish, you can forget the trap. The rabbit snare exists because of the rabbit. Once you've gotten the rabbit, you can forget the snare. Words exist because of meaning. Once you've gotten the meaning, you can forget the words. Where can I find a man who has forgotten words so I can talk with him?

~ Chuang Tzu

First, a disclaimer: Does the label *Ultimate Guide* mean that this book will contain all of the knowledge about permaculture that exists or could exist in the world? Will it answer every question that could ever be asked? No, of course not. It was written by one person in less than one lifetime and cannot possibly contain everything.

This book contains all of the core fundamentals. It whittles down all of the information and theory and experience into a tool that you can use in a practical way.

ULTIMATE | ˈəltəmit |

adjective
• basic or fundamental: *the ultimate constituents of anything that exists are atoms*

Permaculture is a vast field. There are countless courses available to help you increase your knowledge—from introductory weekend classes to full certificate and diploma programs at respected institutions. If you are interested in pursuing permaculture as a career, or if you simply want to dive head first into a bottomless lake of sustainability knowledge, take a course. There you will learn all of the philosophy and theory behind the strategies and skills outlined here. This book, however, was written for everyone else. It is for those who want to begin applying permaculture principles in their own life right away. It is for those who want to become self-reliant, self-sufficient, sustainable, and possibly even develop a community of like-minded people.

I wrote this book while living in a 37-foot bus, traveling from place to place with my husband and three children. However, I am not the proverbial hippie bus mother. I am a geek who likes to experiment with sociological questions, and permaculture is the ideal way to do this. This kind of curiosity spilled over into my own life in the form of our custom-built nomadic home, to which we applied permaculture principles. When you have five or six people living in a

350-square-foot home, inefficiencies and open loops of unsustainability are thrown up into your face as the chaos grows. Chaos from the waste of human living must be managed on a daily basis, and there are no services to truck it away for you. Through this learning process, every square inch of space was analyzed for functionality and impact. There was also the question of making a living while we were nomadic. How does a family survive without anyone having a regular job? This question is eventually asked by all who are interested in sustainability, whether their motivation is to grow food for a living or comes from a desire to simply escape from *the race*. Permaculture taught us to think critically, outside the box. Rather than increasing our income to improve our lifestyle, we dramatically *decreased* our expenses—and our footprint.

At its essence, permaculture is a series of questions. We not only ask where the water comes from, but also where it will go when we are done with it. We ask how

to cool our house without electricity *and* without work. We ask how to grow food abundantly without backbreaking labor or waste. On a greater scale, we ask how we can live as biological organisms within the context of an ecological system, and as humans in a community, without the need to rationalize away guilt caused by knowing our own cost to the environment and to other, less fortunate people.

Not only does permaculture ask the right questions, it has all the answers, encased in a package of design and lifestyle strategies. Up until recent years, permaculture has been heroically promoted and taught by a few groups of ecologically minded people. Many of these people are dedicated legacies of the hippie era, and some are youngsters who are tired of our consumerist society. This anti-establishment culture has had both a positive and a negative impact on the movement as a whole. These hippies and hipsters have protected permaculture and promoted it, albeit within a relatively small minority. They have built up a community that is deeply passionate about the principles behind it. On the other hand, this exclusiveness and protection has slowed the spread of permaculture to the rest of society. If these answers are to be implemented by everyone, the culture surrounding a subculture will inevitably change.

If I have been able to impartially teach the basic principles and how-to aspects of permaculture, I will feel as though I have succeeded. I hope to dust away the cobwebs of preconceived assumptions and ideologies and in so doing further my goal of creating a better, more sustainable world that I can be proud to hand over to my grandchildren.

A Note About Religion

Bill Mollison and David Holmgren, while both equal co-creators of permaculture, disagreed on the spiritual side of things. Mollison, ever the scientist, wanted to distance the official practice of permaculture from any religious features that the community introduced. Holmgren, on the other hand, grew to believe that the spiritual aspects could not be ignored. If you participate in any permaculture design course (PDC), you are likely to be introduced to a variety of quasi-pagan and earth-centric faith practices, which many would argue are essential to a true understanding of permaculture. This book has purposely ignored this element in the quest for practicality and inclusiveness.

1 | What is Permaculture?

Though the problems that face the world are increasingly complex, the solutions remain embarrassingly simple.

~ Bill Mollison

▲ Hiking trail on Mount Wellington, near the University of Tasmania.

A BRIEF HISTORY OF PERMACULTURE

People should think things out fresh and not just accept conventional terms and the conventional way of doing things.

~ Buckminster Fuller

In 1974, young people around the world were asking questions about the way we lived. Hippies, back-to-the-land movements, and communes cropped up all over the world. In Australia, a young David Holmgren was in his final year at the University of Tasmania, and he happened to meet Bill Mollison, a lecturer with a similar interest in ecology and human systems. Inspired by discussions with Bill and experiences in the garden and field sites around Tasmania, David wrote a treatise on how he thought the world should work. It contained the seeds of a groundbreaking design system that managed to combine ecology, human communities, and agriculture into one cohesive whole. The manuscript became part of his graduate thesis, but even more importantly, Bill encouraged David to publish his ideas. The thesis became the book *Permaculture One*, which was released in 1978. The book *Tree Crops: A Permanent Agriculture* by Russell Smith (1924) inspired the word Mollison and Holmgren coined, *permaculture*, but it came to mean much more than just *permanent agriculture*.

Bill Mollison became highly involved in permaculture when the book grew very popular. He eloquently described their philosophy: "[Permaculture] is the harmonious integration of the landscape, people and appropriate technologies, providing food, shelter, energy and other material and non-material needs in a sustainable way."

"Permaculture is a philosophy of working with, rather than against nature; of protracted and thoughtful observation rather than protracted and thoughtless action; of looking at systems in all their functions rather than asking only one yield of them . . . "

—Bill Mollison

Today definitions of permaculture differ. It has been described as a way of life. A culture. A philosophy. At its very core, permaculture is a way of designing all human systems so that they integrate harmoniously with ecology. It grew to include community systems, cultural ideologies, business, art,. . . every facet of human life. What had originally started out as *permanent agriculture* ended up

meaning *permanent culture* because the idea encompassed much more than just agriculture.

David Holmgren defined permaculture as "consciously designed landscapes which mimic the patterns and relationship found in nature, while yielding the abundance of food, fiber and energy for provision of local needs."

One great permaculture teacher, Toby Hemenway, described it even more simply: "Turn every liability into an asset."

After *Permaculture One* was published, Mollison was asked to speak at various educational institutions about he and Holmgren's revolutionary ideas, and he jumped on the opportunity. To his great disappointment, he quickly realized he was only invited to these bastions of learning in order to debate and tear apart permaculture, and he became disillusioned by the university system. In response, he founded the Permaculture Research Institute and an experimental farm in Australia to practice and teach the principles of permaculture. As the movement grew, he designed a course that offered certification. This course has been so popular and mimicked in so many ways that Mollison at one time tried to trademark the word *permaculture* without success, although he did finally succeed in copyrighting the word for educational use. If people want to earn a Permaculture Design Certificate, then they must take the PDC course from someone else who has one.

There was a time when the organization maintained a teacher registry, but

▼ Permaculture is practical but imaginative too.

that is no longer the case. Mollison and Holmgren disagreed on the best way for permaculture education to continue. Mollison wanted to retain complete control of the curriculum and keep it scientific, while Holmgren wanted to allow free reign and the inclusion of religion. In the end, both scenarios played out. Curriculum is controlled, but the delivery of the courses is completely up to the individual, and the community has done what it wants. This makes it sometimes difficult to find an experienced and knowledgeable permaculture teacher, but as Mollison wrote in his book *Travels in Dreams* (1996): "Finally, with hundreds of itinerant teachers turning up anywhere, the system is beyond restraint. Safe at last, and in geometric growth rate—we have won! Permaculture is permanently ungovernable."

PERMACULTURE ETHICS AND PRINCIPLES

The definition of insanity is doing the same thing over and over again and expecting different results.
~ Albert Einstein

Sustainability Has Been Lost

Most of the continents of the world now support many more people than they did a thousand years ago. There was a time in man's history not so very long ago when there were only enough people for the land to support. People ate what grew in their own region and no more than that. For example, the North American continent only supported about one million native people, while today there are more than 300 million people living in the same area.

▲ Cities could avoid smog with proper planning.

When fossil fuels were finally tapped at the beginning of the industrial revolution, an intense population explosion occurred, followed by an even more extreme advancement in technology. Food could come from warm places thousands of miles away and supply fruits and vegetables to hungry people during the winter. Mechanical devices used in farming increased production dramatically. Children survived longer and grew up to have more time to research medical technology, which extended people's lives even further.

Many people will argue about the ethics of having more children, but that is not the issue here. The population is high, but it is still not intolerable or impossible to sustain. The real problem is the manner in which industrialized nations live. All of these people take resources from the land, through farming, mining, forestry, oilfields, and dams, and they change these elements into something usable, by consuming massive amounts of energy. Then, after people use them, the waste is released into the air, thrown in the water, or buried in the land. A bottle of shampoo is made of plastics and chemicals which were mined and grown (using energy and making waste), refined in factories (using even more energy and creating even more waste), then used to wash our hair (where

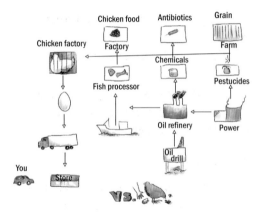

Chicken food   Antibiotics   Grain

Chicken factory   Factory   Farm

Chemicals

Fish processor   Pestucides

Oil refinery   Power

Oil drill

You

Store

VS.

◀ **Energy consumption of factory egg farming versus home laying hens.**

that will never increase or decrease, unless some amazing principle is discovered which magically changes that law. For now, we are stuck with the amount of energy we have. Much of that energy is locked up underground in the form of oil or other materials, and much of it has already been used. It is particularly troubling that most of the people in the world don't use any energy to speak of. The rest of us use it as if it will last forever: inefficiently and wastefully. On top of this we have little care for the *environment*. This has become a dirty word in many circles, but the environment is simply a place where humans live. All other animals care for their own habitat, because polluting it carelessly can make them sick or even kill them. Humans have shortsightedly already done

the chemicals are released into our water supply), and the bottle is most often thrown into a landfill (to which it is transported by a big truck using more fuel and making more smog). Even the so-called "botanical" shampoos didn't come from the renewable fruits of the land but are made using an exhaustible supply of fossil fuels.

There is only so much energy potential available on the earth. It is a static number

▾ **Tornados are becoming more violent and more frequent every year.**

The Ultimate Guide to Permaculture

this, and have thus created a variety of problems, including:

- Climate change. Dramatic changes in temperature and weather, resulting in human deaths, damage to human homes, and destruction of crops.
- Soil degradation. Loss of soil fertility and of precious topsoil due to development and erosion.
- Resource depletion. Running out of resources on which humans depend, like oil or fish.
- Breakdown of human groups. Families and communities losing cohesion and support.
- Increase in addictive behaviors. Cell phones, television, food, shopping, and games replacing relationships and meaningful endeavors.
- Economic and political upheaval. Recession, depression, wars, and oppression.
- National and household debt. Governments and people buying more and more on credit until nothing is worth anything.

In an effort to fix these problems, society has turned to technology, rather than going to the root cause of the issues. We try to fix symptoms rather than considering the events taking place when the resources were gathered and changed in the first place. We use technology to reduce smog, or we build better power plants, not even thinking that this building process itself uses even more energy— energy that simply doesn't exist in sufficient amounts. We're buying energy on credit by using nonrenewable fuels.

People also want to believe that being more conscientious about their environment will change everything.

"If I just ride my bike to work more . . . "

"If I install a water saving shower head . . . "

"But I use fluorescent lights . . . "

The idea that technology or being "good" will save humanity from our problems is a fantasy.

The solution . . .

People in most developed countries are stuck in their high-energy systems, but most of us can agree on some level that this approach is just not working. Whether it is the energy used in our food production, our power sources, or even our political system or economy, the energy we put into it is not sustainable. There are only two outcomes to this scenario. Either everything will continue in its own mediocre way on a gradual downhill slope until it eventually fails, or it will cause a sudden and complete collapse of society. Based on what has been happening in the last decade, mediocrity sounds just as frightening as social collapse.

There is only one clear course of action. When something doesn't work, it's time to try something new. The real solution is a low-energy system. Permaculture makes it possible for humans to switch gracefully to a low-energy system without discomfort. Humans in so-called "civilized" countries could actually make do with 40% less energy without having to sacrifice much. Currently, all of our needs are met

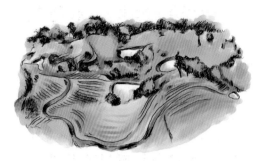

▲ **Working with the terrain and existing ecology is less work and more productive in the long run.**

by energy that came from somewhere else. Our food comes from faraway farms that get their fuel and nutrients from even farther away places, the things we use, the cars we drive—everything comes from far away. A sustainable society is localized. All of the energy used must come from under our own feet.

VALUES OF PERMACULTURE

Ethics is nothing other than reverence for life.

~ *Albert Schweitzer*

Permaculture Ethics

An ethic is like a code of honor, but more specific. Ethics guide our behaviors and are the vehicles by which our destiny manifests itself. There are three ethics of permaculture, and they are fairly simple:

Care for the Earth: All things, living or non-living, have intrinsic worth.

Care for people: Humanity is cared for through self-reliance and community responsibility.

Give away the surplus: The surplus must be shared to fulfill the other two ethics.

In the current system, everything is used once or twice and then thrown away into the water and air, never to be seen again. This is sometimes called a *linear* system because everything makes a straight line from the source to the landfill. Sustainable systems, by contrast, are circular. The used items go back to their source, where they can go through the natural recycling process of the earth and be used again, using very little energy. The same is true of permaculture ethics. When each resource or living creature is valued rather than exploited or destroyed, and people care for themselves as well as their community, an excess of resources is the natural result—and the surplus can then be used to care for the earth and people again.

Principles of Permaculture

Every design system has principles guiding it. For example, in typography, there are a variety of rules that guide the typographer towards good design. A typographer must make the letters appealing to look at, and yet also easy to read. Not only that, but the style of the letters must fit with the overall feel of the document presented and be appropriate to the audience. Permaculture may be a very creative and imaginative method of design and work with some highly variable pieces, but it still follows some basic principles. Different permaculture groups may phrase these principles differently, but the meaning is the same. Most will include twelve principles or more, but I have combined some of these together for simplicity.

1. Every thing is connected to and supported by everything else.
2. Every thing, or *element*, should serve many functions. Students of design

usually learn to make things look nice and be functional at the same time, but permaculture focuses on function alone.

3.   Functional design is sustainable and provides a useful product or surplus. If it doesn't, it creates pollution and work. Pollution is an overabundance of a resource, or something that is simply not used. Work results when one element doesn't help another element.

4.   Permaculture maximizes the useful energy in any system (or, put another way, decreases the waste of energy).

5.   Successful design serves the needs of people and provides many useful connections between elements, or *diversity*.

6.   If there is pollution, then the system goes into chaos.

7.   Societies, systems, and human lives are wasted in disorder and opposition. To stop this vicious cycle, we only use what we can return to the soil and build harmony (cooperation) into the functional organization of a system.

To fulfill these basic principles, permaculture draws from many old and new ways of doing things, many of them considered "fringe." Organic agriculture, alternative building methods, passive solar design, renewable energy, people-powered vehicles, home education, alternative medicine, homebirth, WWOOFing, ethical and socially conscious business, consensus and nonviolent communication, forest gardens, and seed sharing are only a few of the activities that already exist to provide solutions for each of these principles. While many aren't considered "normal" now, most of them were the only way to do things not too long ago, and could be again.

SUSTAINABILITY

To waste, to destroy our natural resources, to skin and exhaust the land instead of using it so as to increase its usefulness, will result in undermining in the days of our children the very prosperity which we ought by right to hand down to them amplified and developed.
~ *Theodore Roosevelt, 1907*

Sustainable Land Goals

When you set out to design a system for a plot of land, the task may seem a bit overwhelming. Sustainability can be a daunting goal, but we can approach it from a logical step-by-step process. The first step is setting achievable goals:

1.   The system should become self-sustaining and productive in the long term. This means that every creature that lives on the land also gets all of its food and personal demands from that land.

2.   About a third of the land should be used for growing food for humans, and the rest should be for animal living space and animal *fodder* (food).

3.   The land should produce more than one needs. The extra can be sold or given away.

4.   A single person can reasonably manage less than 25 acres (10 ha). More than that is too much work, and in fact, the smaller the piece of land area that is needed and used, the better.

5.   The area should be able to provide a full income to the workers that live there. This does not necessarily mean food production—it just means there should be no commute to work.

---

6. Any processing of farm products (like cheese or bread), should be done on site.
7. The beauty of the design should come as a benefit of its functionality.
8. Some areas of the land should remain wild and preserved in their natural beauty.
9. Use low-energy, simple technology.
10. Soil fertility and water quality are your number one concern through every activity that takes place on the land. Those are what keeps you alive.
11. Native species should be used whenever possible. The second option is *exotic* (foreign) species that have been proven to have little impact on your local ecosystem.
12. Use local materials for building projects.
13. Systems should require low maintenance and very little work, and they should take into consideration the culture, society, economy, and legal rights of the people.

How can success be measured?

For the land to be considered sustainable, it needs to produce at least as much or more than it consumes. Rather than measuring success in terms of the pounds of food the land can put out, it is much better to measure the energy that is available in the system and the intangible benefits that are reaped.

1. Water storage should take up 10–20% of the land and should result in greater animal and plant production. Adequate water also creates habitats for ducks (which have their own benefits) and microclimatic changes which ripple out to benefit other elements.
2. Production increases when irrigation is steady and the soil is healthy. Roots can then penetrate deeper and get essential nutrients.
3. Using gravity fed water saves energy, and the setup can be used to recycle water. Electricity usage is a measurable energy (and money) expense, and the less you use it to pump water, the better.

▲ The difference between a "normal" backyard and a permaculture backyard.

4. Tree windbreaks and forests for animal forage should cover 20-30% of the land and will increase production simply by providing shelter and microclimates for plants and animals. They will also provide food and homes for predators that eat pests.

5. Natural areas and wildlife corridors conserve water by trapping condensation and provide usable timber, materials and shelter for wildlife.

There are also non-tangible ways of measuring success:

1. When the farmer, despite *not* having a tractor, is able to perform less physical labor because of the interconnectedness of the system.

2. When the land has recreational value for people to enjoy.

3. When future generations can enjoy the land.

GETTING STARTED

There is an orderliness in the universe, there is an unalterable law governing everything and every being that exists or lives.

~ *Gandhi*

Patterns

Almost all of the design strategies in permaculture are directly drawn from the recurring patterns found in nature and mimic the ongoing processes of the natural world. The wilderness may seem chaotic, but in reality there is an ordered method to everything, from the physical structure of organisms to the invisible chemical cycles that keep them alive. Water, nitrogen, seasons, birth, and death are all part of the patterns we live within as human organisms, whether we are aware of them or not. We must become intensely observant of the natural world and clever enough to use what we discover. Cycles exist in both space and time, taking up area on land and spanning months of the year. The growth of life on a piece of land is also not flat but vertical as well.

Patterns in nature also repeat. The Fibonacci series of numbers is the most well-known pattern, a mathematical sequence that creates a proportional spiral found in everything from snail shells and sunflower heads to the leaf patterns of oak trees. Intertwining coils are found in the stems of plants and the structure of our

▼ Mathematical patterns and formulas exist behind every natural structure.

own DNA. There are predictable patterns in the way that birds flock and fish school together. These patterns exist in nature because *they work*. The life we enjoy today would not exist without them.

What is ecological succession?

When an area of ground is cleared of plants in the natural world, without interference from humans (for example, as a result of fire or storms), it begins to repopulate itself through a series of stages. We want to speed up these stages or *succession* and use them for our own

▲ Even a fern follows the spiral pattern.

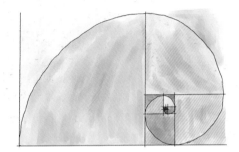

▲ Fibonacci spiral based on sequence of numbers: 1, 1, 2, 3, 5, 8, 13, 21, and 34.

benefit. This saves time and energy for humans and benefits the ecology of the environment because one is still using the natural cycles. These stages are:

Stage 1. Broad-leaved plants that spread very quickly to cover and protect the soil, sometimes called *pioneers*. Usually

This area experienced a forest fire that cleared the vegetation. Pioneer plants have moved in.

The Ultimate Guide to Permaculture

these are thorny and inedible, like thistles, and don't live very long. Others are thorny and edible, like blackberries. These also tend to add important minerals to the soil.

Stage 2. After the broad-leafed plants, herbs and shrubs begin to grow, some of which are edible. These last longer. They compete with each other and the Stage 1 plants for resources like water or light, and they eventually win out over the pioneer plants.

Stage 3. Trees begin to grow when the shrubs and herbs have created an environment that is beneficial to them. Some of these trees will bear edible fruit. They will begin to shade the pioneer plants, which die and fertilize the soil.

What ecological succession is not . . .

Companion planting is not the same as ecological succession. For example, if you plant clover under a fruit tree thinking that you are stacking the stages and adding nutrients to the soil, you are actually inhibiting the soil from absorbing water and thus can kill the tree. Putting plants from Stage 1 with plants from Stage 3 is not always a good strategy, and not every species works out. Ecological succession is a process by which bad and cleared soil fixes itself through certain species of plants that grow one after the other, or *succeed* each other. This happens over a period of years or even decades. Companion planting is a system of growing plants close to each other that happen to have beneficial properties for each other, over a single growing season.

The Importance of the Edge

Squares and rectangles are very rare in nature. Instead, we find spheres,

▲ The edge of a bioregion has more organisms than the center.

circles, cylinders, and especially spirals. From the tiniest perfect shell no bigger than a fingernail to massive hurricanes and awesome galaxies, the spiral is the preferred shape of the universe. The edge of a forest gets more light and nutrients than the center, and because of this the edge is much more productive. Even more important is the relationship between edges. The line between the ocean and the shore, or the mountain and the valley, is as diverse as the two areas on either side. To take advantage of this *edge effect*, we use circles, spirals, curves, vertical growing space like trellises, and zigzags to create more edge space. The space between the orchard and the chickens or the house and the garden is designed to maximize the potential production power of the additional light and diversity.

Types of edges:

Spiral: An herb spiral is five and a half feet (1.6 meters) across and shaped like a snail shell with a ramp leading up to the center. Herbs that like sunshine are planted on the sunny side, and herbs that prefer shade are planted on the other side. An herb spiral is often placed in the kitchen garden.

Lobular: *Lobular* describes a shape that is roundish and hangs out, like an ear lobe. It is an edge that is put alongside a pathway or the side of a garden.

Chinampa: This ancient system was used in Mexico for thousands of years. Chinampas are land banks built up between ditches of water. Fish are raised in the water, which fertilizes the plants, and the plants also have easy access to the water because they are so close to the ditch. The fertilizer from the bottom of the ditch is periodically brought up to the garden beds.

Strip cropping: Several types of plants are grown in strips next to each

▲ Spiral herb garden.

▲ Lobular garden.

▼ Trellising with a shade garden underneath.

The Ultimate Guide to Permaculture

▲ Chinampa of Central America from *L'Illustration, Journal Universel,* Paris, 1860.

other for mutual benefit. This doesn't mean straight lines but rather curves, zigzags, keyholes, or following the contours of the land. Strips can also follow a pattern.

Permaculture words to know . . .

Polyculture: Monoculture is the type of picturesque farming that one typically sees in North America, with long rows of a single crop grown over a large flat area. The "amber waves of grain" in America is monoculture, the complete opposite of polyculture. Polyculture exists when a variety of plant and animal species are mixed together for mutual benefit. Orchards can be clumped closely together, rather than spread out in neat rows, and have herbs and ducks under them. A climbing plant can be grown with a tall plant, such as corn.

Aquaculture: Water systems have the potential to produce much more protein per square foot than an equivalent area of land. A successful aquaculture system is patterned after productive land-water edges such as swamps and coral reefs. While some aquaculture systems are simply an aboveground tank with plants growing off the dredged fertilizer and look very similar to hydroponic systems, in a permaculture design aquaculture looks more like a natural pond. The pond would contain fish and have a very curvy edge with plants and animals thriving together.

Elements: Any feature on a piece of land is an *element*, whether you intentionally placed it there or not. This could be a clump of trees or herbs, a pond, or a pile of rocks. These elements are part of an overall design, each one thoughtfully used to make the land more productive. Each element is also connected to everything else in as many mutually beneficial relationships as possible. This wide variety of connections, or *diversity*, increases efficiency and also places value on elements that formerly might have been seen as annoyances. Every "problem" can be turned into an advantage. For example, weeds are simply Stage 1 plants (or pioneers) and can be turned into mulch. A big immovable rock can be used as part of the supporting wall of a house.

Elements as goals:

Before you move on to the next phase of design, you should first think about your goals. What do you want the land to do for you? What do you need to live? Write these goals down and turn them into elements:

A place to live =	a house
A place for friends and family to stay =	a guest house
A way of growing food in winter =	a greenhouse
A solution to drought and irrigation =	water storage
A way to stay organized =	a tool shed
A place for animals to live =	a barn
A place to grow food =	a garden
A place for food and yard waste =	a compost pile
A place to store fuel =	a wood pile
A source of protein and manure =	a chicken coop
A source of fuel =	a woodlot
A way to stop the winds =	a windbreak
A way to control water flow =	a dam

▾ **Self-reliant cabin with living roof.**

Inventory of elements:

1. Now that you have a list of goals, or elements that you *want* to have, you also need to take an inventory of the elements you already *do* have. Any characteristic of your land, whether you see it as positive or negative, must be written down. This includes large rocks, hills, marshy places, existing structures, trees, etc. Of course, to create a map like this you will have to carefully inspect every corner of the land and use every sense you have to observe the finer details. This may mean taking notes for an extended period of time. Through your own observation, record:
   - Temperature changes from one area to the next
   - How much effort it takes to go from one place to another
   - Where prickly plants tend to grow
   - Where insects seem to group or swarm together
   - Where different species of trees are growing and in what conditions
   - How water moves across the land when it rains or snow melts
   - Where there have been fires (even long ago)
   - Which trees have been shaped by the wind, and the direction the wind comes from
   - Where the sun is warmest and where shadows move
   - Signs of animals moving, eating, and sleeping
   - Signs of sources of groundwater, such as deep-rooted trees
   - Plants that produce fruit before other plants
   - Poisonous or unpleasant plants
   - Erosion or ditches caused by erosion
   - Damp or boggy ground, which may yield peat or clay
   - Dead wood or logs that can be harvested
   - Slopes and hills, their height, and which sides are sunnier
   - Cliffs, rocky places, and rough areas

2. Categorize all of the resources you have mapped and list them into three groups: life, energy, and social. Life resources are the plants, animals, and insects growing there. Energy is the potential wind, wood, water, or gas energy you can utilize. Social resources are the teaching, recreation, and gathering possibilities for people. There are also resources off the land, such as restaurants and markets for selling products, sawdust from sawmills, schools, and a population of people who might be a potential market for your farm produce.

3. Analyze each element. Think about what it needs to live and what it needs to be useful. What are its characteristics, its behavior? What does it produce and what beneficial functions does it provide?

4. Categorize these by input, characteristics, and output. Input is what the element needs to function; anything that allows it to be useful or keeps it alive. Output is what it can provide to other elements. Characteristics are the attributes that make that element what it is. Make sure you number the element for reference later. It is easiest to write each element on a 3 x 5 card:

1.	Rock Pile
Input	Sunlight (to radiate heat)
	Human labor to move the rocks around
Characteristics	Dark slate
Output	Insulating and passive solar properties
	Structural support
	Roofing, flooring, wall for passive solar heat
	Windbreak, shelter

2.	Herb Garden
Input	Sun
	Water
	Mulch
Characteristics	Variety of perennial herbs (comes back every year)
Output	Medicinal herbs
	Cooking herbs
	Bee forage

3.	Chickens
Input	Food
	Water
	Grit
	Shelter
	Other chickens
Characteristics	Barred rock, dual-purpose breed for both eggs and meat
Output	Eggs
	Meat
	Feathers
	Manure
	Methane
	Foraging

4.	Deer
Input	Food
	Water
	Forest
	Other deer
	Marked territory/path
Characteristics	White-tailed deer
Output	Meat
	Manure
	Methane
	Foraging

More examples of outputs:

Pond: irrigation, animal water, aquatic plants, fire control, bird or fish habitat, passive light reflection, firebreak, bamboo garden.

Plants: windbreak, food, privacy, fuel, trellising, erosion and fire control, wildlife habitat, mulch, microclimate buffer, food, soil improvement, shelter, warmth, friendly insects.

5. Once you have taken an inventory of all of the elements you have and that you plant to build, you will need to match the outputs with the inputs of other elements. For example, if bees need to eat and the herbs can provide food for them, they should be planted in proximity to the bees. The chickens need food every day, and so the coop should not be too far from the house. At the same time they produce manure, and so it should be near the

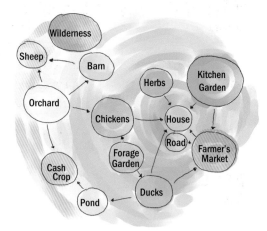

manure pile. To save space and work, the chickens can be placed in a side room of the barn, for easy access. The woodlot can provide forage for chickens, and so their pen could open out into the trees. Lay the cards out next to each other on the table so you can visualize these connections.

6. If you can't leave the cards undisturbed as you go on to the next phase of design, draw a flowchart of how the elements are organized and connected.

THE ZONES

Much good work is lost for the lack of a little more.

~ Edward H. Harriman

The easiest way to start the design process is by dividing the land into *zones.* Zones are areas classified by how much human intervention is needed to maintain them and are placed according to their distance from the center of human activity.

Zone 0 is the house (or on a larger scale, a business or village). This is where you live, but it is also very connected to the garden and the greenhouse. This is where you will process food from the garden, host

▾ **A gulley providing irrigation for climbing plants and trees.**

dinner parties, and produce human waste. You will spend the most time here, and so everything else branches out from this zone. Things that you do very frequently need to be near the house, and things that you do only once a week or once in a season should be farther away. These elements include the kitchen, shade room, greenhouse, trellises and their attached vines, composting toilet, and house pets.

Zone 1 is the area immediately surrounding the house, although it may extend down walkways or driveways into other zones if you use those places very often. This is where you will grow plants very intensively in a sheet-mulched garden. It may also have a small pond, various outbuildings that you use frequently (such as a tool shed), small fruit trees, and a low windbreak. The trees here are dwarf varieties, or perhaps you might have multiple varieties grafted onto one tree. Water in this zone comes from a well or rainwater tank and is fully controlled with pipes and hoses. This area should be able to produce most of the food a family needs to live.

Zone 2 is the area that extends slightly beyond the gardens surrounding the house. It has plants in beds like Zone 1, but these are bigger and used for major crops that make up the staples of your diet (like grains or potatoes). Worms, rabbits, chickens, ducks, or fish can be raised in a small yard, just beyond the proximity of the Zone 1 gardens. This is also where you would keep fruit and nut trees and the compost heap and use hedges and trellising to maximize the edges. In a suburban area, this is as far as the land would extend, although you could incorporate a very small Zone 5 into one corner. On a larger piece of land, you might also keep a few goats here, or a single milk cow. The zones are not classified by what is in them, but by how close they are to your house.

Well-mulched pathways of a kitchen garden.

▲ Most of Zone 2 is mulched with straw and other materials.

Zone 3 is for rural areas and properties that are larger. This is where you would have a larger, unpruned orchard and other trees that act as a windbreak for the house. The goat pen and beehives are placed here, and the area is full of living mulches, plants for animal fodder, and firebreaks. Water is stored in the soil

▲ Mulched beds with drip irrigation and cover crops.

in swales, or it is caught with small dams and sent through ditches rather than pipes. The barn would be located here, and this is where you would raise cash crops and animals that are going to be sold for profit.

Zone 4 shows the benefit of having a large piece of rural land. This is an area of long-term development through a woodlot, dam, and extensive tree planting. Windmills and large animal stock are placed here, along with large-scale water harvesting. Pigs work well in a forest zone like this. No mulches are used, and hardy edible plants are foraged from the edges of the forest where they need very little ongoing care. Water is managed with small dams, rivers, and windmill pumps for irrigation into Zone 3.

Zone 5 can, and should, be placed on any sized property. This is a wilderness area, the nature preserve. This is where wildlife corridors and forest growth can be fostered. This is where you can observe the untamed wilderness and have fun in the woods. In an urban backyard, this would simply be a back corner that you leave to the birds and bunnies.

Creating zones and sectors:

1. Now that you have taken an inventory of all the assets and elements of your land and how they can benefit each other, you can place them in zones. Putting the elements in the right zone saves human energy, which is essential to being productive and efficient. First you need a detailed map of the property, which should show water, major existing plants, geological formations, roads, paths, fences, structures, and power sources.

2. Using the "bull's-eye" of the zone as a template, draw lines on your map designating your zones. They shouldn't be perfect circles but should follow the contours of the land and its various characteristics. It is a good idea to color code everything on the map for easier visualization.

3. Positioning the house is largely dependent on access. Materials must be brought in to build the house, and usually you would need to be able to bring a large truck in to transport

Element	Zone 1	Zone 2	Zone 3	Zone 4
Garden	Totally sheet mulched	Some mulch and protected trees	Cover crops and green mulch	Cover crops
Trees	Very controlled, dwarf varieties, grafted trees	Pyramid pruning, grafted trees	Unpruned, seedlings used for grafts	Thinned seedlings, unpruned
Water	Rainwater collection, wind pump, irrigation pipes	Well or earth reservoir	Water storage pond, fire control	Dam, river, swales
Structure	House, greenhouse	Barn, poultry sheds	Feed store, animal shelter in fields	Hedge, woodlot

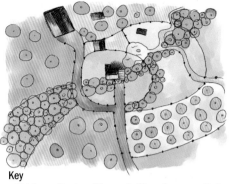

Key

□ Chicken □ Duck □ Pig □ Cattle □ Crops □ Orchard

those materials, and a road must be built that can accommodate the large truck. You probably have a car that you will want to park near the house in a driveway, and you will have to maintain that driveway. It's a good idea to build the house close to the main road rather than set back away from it, so that the driveway is shorter. It is cheaper and easier to have a short driveway, and you will be less isolated.

4. If you have the option of positioning a house that you will build in the future:

- Build it halfway up a slope, rather than at the top or at the bottom (ideally, water storage can be located above the house on the slope, to be gravity fed down to the house). At the top it will be buffeted by winds, and at the bottom you will be fighting vegetation and waterlogging.
- If possible, choose the side of a slope with the most exposure to the sun, unless you are in an extremely warm place, in which case the side with the cool breezes should be chosen instead.
- The house should be very close to the supply of power, whether it is your own wind or solar, or city power. Wiring and piping electricity any distance is difficult and expensive.
- While any soil may be improved, avoid putting the house on any very good soil. You should, however, make

▼ **Barn and well set uphill from the house.**

sure the soil has good drainage by digging a hole three feet deep and filling it with water. If it doesn't drain at all within one minute, there's a problem, and your house will have moisture and flooding issues. You will have to either improve the drainage or find another spot.

- Noise, pollution, and privacy may require that the house be farther back from the road or away from the neighbors. These are important factors and worth making the driveway a little bit longer, because you don't have control over the highway or the people next door. But you can still build up an embankment and grow hedges to provide noise control and privacy.

- The view is not a good enough reason to put a house somewhere. Sacrifice the view for all of the above factors and build a small outdoor guest area somewhere else for a better view. If you really want a view from your house, build a tall house.

- One serious, but maybe not so obvious, issue is disaster prevention. Don't build the house on a floodplain, or on a slope so steep that you may become the victim of a landslide, or on an eroding beach, or near a rising sea or an active volcano. In a tsunami area, build above the reach of even the worst tidal wave. In tornado and hurricane areas, look at the past weather patterns and find out if you are in the path of these storms. Is there a sheltered area on your land that works better?

4. Refer to your map to arrange the elements into zones, beginning with Zone 1. You can start off by simply

arranging the cards into piles, based on your map and the definitions of what works best in each zone.

5. Now you can divide the zones into *sectors*. Sectors are the outside influences (sometimes called *energies*) that have an effect on the land and how it is used. You already inventoried many of these when you categorized your elements. Sectors include:
   - The direction the wind blows and where it is the windiest.
   - The amount of rain that falls and where it rains the most.
   - The direction of the sun and how long it shines in all seasons.
   - Prying neighbors.
   - A beautiful view to preserve, or an ugly view to be blocked.
   - Swampy areas and places that flood.
   - Hills and slopes.
   - The angle of the sun in all seasons.
   - The location and proximity of the greatest fire danger.

6. You can outline the sectors on the map of your property like this:

Arranging the elements:

1. Since you already numbered each element on your 3 x 5 cards, you can

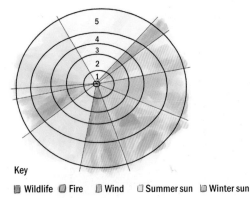

Key

▢ Wildlife ▢ Fire ▢ Wind ▢ Summer sun ▢ Winter sun

▲ Sectors intersect the zones.

The Ultimate Guide to Permaculture

draw a circle with a number on the map for each one. You will need to keep in mind where they will have the most beneficial relationships to all of the other elements, which you already thought about when you arranged them in piles or drew your flowchart. For example, if you are placing a large tank or pond for raising fish in Zone 2, it needs to be near a water source, which might be in Zone 1. You would also want to put it somewhere with easy access to the house (where you would process the fish and eat it) and also near a compost or manure pile to collect the manure. The best spot would then be on the edge of Zone 2, backing the Zone 1 garden to offer heat reflection and next to the compost pile.

2. Keep in mind the sectors. If a place is windy, you will add a windbreak there. If it gets flooded, add trees that tolerate lots of water. If the house is slammed by rain, add more trees next to the house. Put the garden where there will be the right amount of sunlight.

3. Small roads and trails should be worked into the design now, rather than trying to figure out access to more remote areas later. The road leading immediately to the house from the highway should be sloping uphill slightly, even if you have to build the grade yourself. This is to drain water away, and the slope gives the road better access to the sun in a wet or snowy climate. On the other hand, roads should never be built with steep slopes but should follow the contours of the land. The road can be part of a swale or dam, with rainwater runoff benefitting either structure, but even if it can't drain the water directly into a swale or dam, it should drain into a ditch with a pipe going under the road and diverting water to the right place.

4. Zone 2 may have chickens and an orchard and should have heavy duty wire mesh fencing and possibly barbed wire, electric wire, and thorny shrubs. A thick hedge is an ideal goal to work towards, because it will keep out most animals and needs less maintenance than a wire fence. Use hedge species that are appropriate for your area and will benefit the most inhabitants of the land by feeding chickens or bees and providing homes for birds.

5. If you hadn't noticed in the previous steps, slope is the deciding factor for the placement of most of your elements. Water flow, passive solar heating, solar photovoltaic panels, exposure to the weather, the view, the angle of the sunshine, the effects of erosion, all of these are determined by the slope. Use the slope as an asset in each phase of planning. The top of a slope is in the greatest danger of erosion but is also the best place for catching and storing water. The middle of a slope is the best place for a house, since it is the most sheltered, and the bottom of a slope is perfect for gardens and animals. Flat areas at the bottom are easily converted to water storage but can also be prone to high salt levels through evaporation.

6. When you get to Zone 5 in your planning, you won't be doing anything with it. You might need to remove any man-made problems in that zone, but once you are done, you will simply fence it off and leave it alone. All of your zones would return to their natural state when left alone, and Zone 5 is no different. Planting native species is the same as planting a garden, which misses the point of what Zone 5 is supposed to be.

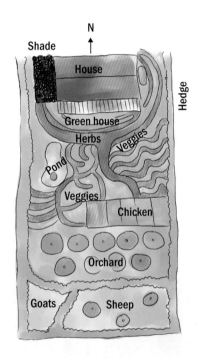

These design examples work with the terrain and crowd elements as closely as possible.

CLIMATE AND MICROCLIMATES

Sunshine is delicious, rain is refreshing, wind braces us up, snow is exhilarating; there is really no such thing as bad weather, only different kinds of good weather.

~John Ruskin

Climate as an Asset

Weather is one of the top complaints of all gardeners: too much sun, too little sun, too much wind, or too much rain. Conditions are never perfect. North America has almost every climate represented from the driest places in the world to full tropical rainforests, which not only increases the difficulty in growing plants but also provides tremendous opportunity to grow just about anything.

Most people cut a square out of their backyard and plant perfect rows of the same varieties of species that you might find at a grocery store, which is a very narrow view of gardening. Not surprisingly, they often meet with failure and give up. Not only are those species developed for a specific type of commercial farming, but the chances that those species were designed for your climate and weather are very slim.

The real extremes of weather and climate can be used and defended with a little planning. The importance of choosing the right species is emphasized throughout this book, and it starts with climate. A hardy, heritage species developed by home gardeners specifically for your climate is likely to do better, obviously. You must set up microclimates using trellises and water diversion, either to retain moisture and coolness in the desert or to reflect heat and store water in a colder climate. The key to this is remembering that weather isn't a problem; it is an asset to be used.

Climate Zones

Most people are familiar with the USDA Plant Hardiness Zone Map. These zones are

▾ **This trellising creates a shady, cool microclimate in an otherwise hot, sunny area.**

1. Some plants require frost to produce fruit, and frost also adds water to the soil. It's just not possible to stop frost everywhere on your land, and nor would you want to. If you are situated on a slope or hill, you can locate the Zone 1 or 2 elements above the frost line (an area known as a *thermal belt*). This was discussed briefly in the previous section about placing your house. The reason this is important is because cold air settles on the tops of hills, causing frost, and also flows downward, pooling in valleys. This leaves the thermal belt a little warmer.

2. A wall built with dark stones absorbs heat and can prevent frost by radiating heat at night. Plants will grow faster next to a wall like this, but a light-colored wall will assist some sun-loving plants to ripen, so use whichever color works best for plants that will be growing there.

3. A large body of water will warm and cool more slowly than the surrounding environment and thus serves to modify the temperature of its immediate vicinity. This creates a small microclimate, and this is why there is less frost near the ocean.

4. Trees insulate the earth and trap heat in, preventing frost. One of the easiest ways to protect your gardens from frost is to create a canopy of trees surrounding a clearing no wider than half the height of the trees. The clearing allows a little bit of sun in, and the ring of trees around it will keep the frost out. This must be built up over time, by first planting a quick-growing fruit variety which can shelter a frost-hardy legume species. When the frost-hardy canopy has grown, you can trim back

▲ **Tree canopy insulates a clearing, protecting it from frost.**

the other trees. This canopy also acts as a rain gutter, directing water to the plants below. In very cold climates, use evergreen varieties to provide *biomass* (which holds in heat), and in very warm climates use light-colored and shiny trees to reflect the heat.

Wind:

Too much wind can harm wind-sensitive plants, blow away seeds, lower the temperature of the soil, dry up moisture, kill young animals, and make work and life intolerable. Winds of 15 mph, which in some places is the average wind speed, are strong enough to reduce production. Winds of 20 mph are enough to cause physical damage to plants. A windbreak can either block winds or channel them in the direction you want them to go. It can also ensure that plants produce and animals gain weight. Windbreaks are made of almost anything, can be edible for animals and bees, and can provide a home for beneficial birds. Near the ocean, a windbreak is the first priority in establishing an orchard.

1. When planning your sectors, take note of the direction of the prevailing wind or

▼ **Strategic tree and species placement makes a good windblock.**

Successful natural windbreaks:
- Choose pioneer species that are easy to grow.
- Plants should have fibrous stems, like palms.
- Plants should have fleshy, fuzzy, or needle leaves.
- Use an earth mound or tall solid fence to protect the plants as they grow.

the direction the wind blows most often. You can do this by tying a streamer or flag on a tree or stake. Plant windbreaks all over the place, but make sure you take advantage of the wind for energy generation, either with a turbine for electricity or by pumping water.

2. Because cool air always goes downhill, a slope can create a wind that sweeps through a valley. In some large valleys, the wind will flow uphill during the day and downhill at night. The side of a hill with the prevailing wind will experience faster wind speeds going uphill, and as the wind reaches the other side it

will be disrupted and slow down as it goes down again. A similar effect happens near large bodies of water: During the day, warm air rising creates a breeze that circles towards the land, and at night as the air cools, it reverses direction and circles towards the water.

3. A small shelter can be built around plants to protect them from wind, such as an old bag wrapped around stakes, a metal drum, old tires, or straw bales. Traditional structures like *cold frames* (wood frames with a glass top), *cloches* (capes or bells that sit over the plant), and milk jugs with the bottoms

▾ **Tiny bamboo shade screens for fragile seedlings.**

cut off can all work. A building can be protected with bushes and vines planted around it. Even snow or dirt piled up can be an effective insulator.

4. When choosing the material of your windbreak, remember that trees are both a benefit and a competitor. They provide firewood, stop erosion, create privacy, and are a habitat for animals. They also have large roots that need lots of water, and windbreak trees won't give you much fruit. The best way to start a windbreak is to plant fast growing species of trees and shrubs mixed in with some slower growing trees. The slower growing hardwood trees will live longer but the quicker growing ones will do the job as others mature.

5. In coastal areas, choose tree species with rough bark. They should also either be hardy pine or have very thick leaves which hold in moisture. The easiest way to choose is to see what is already growing in your area.

6. The quickest way to shelter the Zone 1 garden from wind is to build a trellis extending from the corners of the house and plant climbing vines that grow up and cover them. These will create a nice living space around the house, insulate structures, control the wind flow, trap the sun, and provide something to eat. These grow fast enough to provide wind shelter as trees are growing, but make sure to pick the right species as they can quickly take over and embed themselves into structures permanently.

7. Keep in mind that many people run into a serious and dangerous problem because of their tree windbreak. In rural, forested regions, you might create a clearing in the trees to build your house, which effectively shelters you from the wind. However, because the trees were sheltered by each other, their root systems are weak and their new exposure to the wind may be

▾ Trellising holds up a kiwi tree and provides a garden space for shade-loving plants.

enough to knock them over onto your new house.

Temperature:

Every 330 feet (100 meters) of altitude away from sea level is equivalent to a temperature change of 1° of latitude from the equator. For example, if you are hiking exactly on the equator, and you climb to a height of 1,650 feet, the temperature will be the same as it would be 5° from the equator. This is why a tropical region can have snow in the mountains. A body of water also modifies the temperature of the surrounding air through evaporation. In very hot areas, even a small body of water, or a fountain, can cool the surrounding area. There are various means of modifying the temperature:

1. The most obvious way to create a warm microclimate is with a greenhouse, which is particularly valuable in the winter.
2. An earth mound can be built near the house on the west side, protecting the house (and the garden on the other side) from the hot evening sun. It can also provide insulation, because earth will slow down temperature changes.
3. In the spring, remove mulch from the soil in your intensive growing areas that the dirt can warm up, because mulch doesn't have any heat conducting properties. Mulch (and living ground cover) should be used at almost every other time in the Zone 1 gardens to help the soil retain moisture, reduce erosion, maintain a stable temperature, and stop weeds.
4. Plants release water vapor, cooling the air and causing humidity. Filling a porous earthenware pot with water and covering the top with a heavy, wet fabric can create a similar effect. Place these around in the area you want to cool.
5. Plants absorb energy from the sun, and forests are tremendously efficient at it. The canopy absorbs the heat during the day, shading the forest floor, and cool air is drawn in. At night, warm air flows out. This effect should be enjoyed and used through planting forest gardens and placing animals within it.
6. Blocking the sun is a very quick way of cooling something down. A very thick hedge on the west side of a house blocks the hottest sun and also blocks the wind. A tree planted on the sunny side of the house shades the wall and roof in the summer but allows the sun to shine in the winter. One strategy is to plant shiny trees (like poplar) in an arc around the house or orchard, facing the sun. If planted on a slope, this has the

▾ This house isn't underground, but has an earth mound to insulate it.

▾ A ring of deciduous trees reflect light and heat back onto a pasture.

additional benefit of warming the cool air pooling at the bottom.

Light:

1. Plants need light, and redirecting sunlight can mean the difference between green tomatoes and red ones. Any plants that have a high need for light and warmth can be placed on the sunny side of banks or on a small hill behind a pond.

2. Trees with light-colored leaves can be used to reflect heat and light. A sun-facing wall can be used in a similar way, and it can reflect the sun in winter. Plants will ripen faster and more completely if they have a reflective surface behind them.

3. A temporary shelter made out of fabric or other material can be used to provide shade and prevent sunburn during the hottest parts of summer.

PUTTING IT ALL TOGETHER

When planning for a year, plant corn. When planning for a decade, plant trees. When planning for life, train and educate people.

~ Chinese proverb

Planning Ahead for the Unexpected

Nature has its own way of disrupting our plans. The only way to prepare for this is to plan ahead for extreme climate changes and natural disasters.

Fire: Fire usually comes from one direction, dictated by the topography of the land and the general wind flow. This is why planning a fire sector is so important. There

are no guarantees, however, so if fire is a common local disaster, protect all of your vital elements no matter where they are located on the property.

The fire sector should be kept clear of litter and dead logs, and grass should be kept short. The pond, or any element (or any animal grazing area) that keeps the earth bare of vegetation, such as a road or rock wall, can be placed between the fire danger and the house. A windbreak of fruit trees or willows (which aren't as flammable) should be planted at the bottom of hills since fire travels faster uphill. Avoid pine trees, which catch fire and leave flammable litter everywhere. By creating a series of bare ground and deciduous tree barriers, you can slow or even stop fire from reaching the house.

The house is the most valuable item on your property, and so extra precautions should be taken. Build a brick or stone walkway around the house at least three feet wide. Keep it clear of debris or even a doormat. Use window frames and screens and a metal roof. If you are very concerned, build sprinklers around the house with their own supply of water that runs without electricity.

Earthquake: A single level home is safer, but the real secret to an earthquake-

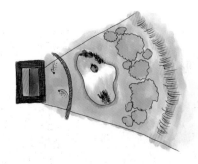

▲ **Fire zone with deciduous trees, swales, water, and brick.**

proof house is the materials with which it is built. Steel beam construction bolted into the ground is more likely to stay standing but may not be feasible for many people. The alternative is flexible construction made of wood or bamboo that is bolted securely to the foundation. The foundation should be steel reinforced concrete (or *rebar*). Every home, whether you built it yourself or not, should have one central room designated as a safe room that will stay standing if the rest of the house were to collapse. This room should not have any windows or heavy things hanging on the walls, and it would be quite small in order to withstand the force of an earthquake and the shifting of the house. Usually a closet or bathroom can serve this purpose. It is a good idea to store nonperishable food and jugs of water in this room.

Flood: Be informed about the history of the flood level and learn the history of flooding in your area. Be smart and don't put your house in a floodplain, but also avoid a steep slope that can become a mudslide during severe rain. If you find that you do live in a floodplain already, build your house on stilts even if people think you are crazy. An existing house can be protected from flooding by using dams, swales, and runoffs to effectively

direct excess water away from the house. Extreme flooding can be controlled by the use of dykes made of sandbags. It is also important to prevent contamination of drinking water by sewage. If you have a composting toilet as this book recommends, that is not a big concern, although a multitude of composts and fertilizers will compromise the water supply should they flood. Electrical systems should be set far above the flood line, and there should be an easy cut-off switch to turn everything off quickly. Propane tanks are an additional hazard as they float and crash into things, and so they must be securely anchored.

Hurricane & tornado: Buildings should be flexible, as for earthquakes, and the roof should have a very sharp 45° angle. The wind force will be more likely to push the building down, rather than pick it up. A bamboo windbreak can prevent major damage if planted on the side of the house where there is usually the most wind. It is also a wise idea to plant a small backup garden in a sheltered place, away from the Zone 1 garden, so if things go wrong, you will still have some food and seed stock. Even more importantly, a storm cellar or safe room should be built to escape into and to store food. A safe room is for people who can't build an underground cellar, and it is a small room built of stone or concrete

▾ **A house built on stilts in a floodplain.**

▾ **Example of a storm cellar**

▲ Snow creates mud instantly when it melts.

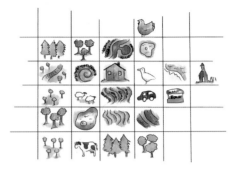

▲ Drawing a grid of the elements in relation to each other can help you check your positioning.

in the center of the house with a solid door installed with a drop-down bar lock. A storm cellar is an underground room, either in the basement with an inside door or built separate from the house with an outside door. A root cellar often doubles as a storm cellar.

Snow: While snow is an inconvenient and dangerous winter hazards that causes power outages, driving accidents, and isolation, the biggest problem around your house will be roof damage and flooding. The snow melts and tends to pool on your roof and Zone 1 areas. The worst damage happens when snow melts on your roof on a warm day, only to freeze again at night or during another cold front. The water forms a pool at the edge of an older roof, or drips down to form large icicles, which freezes again and expands. This causes stress to your roof and eaves and can create cracks in your walls. Prevent this by using a metal roof and making sure the slope is steep enough for snow to slide completely off. Ground surface flooding can be prevented with swales and drainage ditches leading to the pond.

How do I double-check my positioning?

Placement of elements is always relative. Where is this element in relation to the house in Zone 1? Where is it in relation to the wind sector? Where is the chicken coop in relation to the other animals? How will it benefit them? A chicken coop is placed in Zone 2 but bordering Zone 1 very closely because you must go there every day to collect the eggs. It should be placed away from the fire hazard sector, but because chickens pick the ground clean, the pen itself can be used as a firewall between the fire sector and the house. It should be next to the vegetable garden so you can easily move the manure, but it should also connect to a forage garden where the chicken will eat. When starting out, it is important to crowd everything in as much as possible. Cram plant life up next to the house and stick the chicken coop as close as you can to the house. Zone 2 might only be fifty feet away, because this saves you time and labor. It is easier to thin things out than wait for

> **6 rules of design**
> - Is it the appropriate size?
> - Is it in the right order?
> - Does it minimize wasted space?
> - Does it utilize the edge effect and communities?
> - Does it use simple and easy species?
> - Does it use appropriate patterns?

them to grow, and you can branch off to the farther zones later. Each element is placed in relation to everything else and to you—next to, close by, behind, away from.

Besides the zones and an element's proximity to something else, which saves your own energy, you should also consider an element's energy consumption and needs. Your goal is to stick them where they will generate energy or at least conserve energy, rather than consume it. For example, a greenhouse should be up against the house not only to radiate heat into the house but to absorb it during the winter as well. A windbreak should be a fast-growing tree or shrub if possible, rather than a rock wall or embankment, because it takes less work to grow a tree than to build a wall.

Off to a running start:

It is pretty obvious that the ultimate goal of permaculture is to build up a system that relies on plants and animals, but mechanical and non-natural devices sometimes come in handy for certain tasks in the beginning. A solar panel isn't made of sustainable materials, but it is still a low-energy and necessary part of building up a relatively sustainable system. It is good to use a backhoe once to put in a pond and get things going quickly. You will want to prioritize what you are going to focus on in the beginning and divide the tasks into phases:

Phase 1—Place the most vital elements. This includes access, water, and power. You will probably not be able to implement all of your power-generating dreams at once, but install the most necessary power supply, even if it comes from the city. Build road access and implement your water collection and storage system designs.

Phase 2—Build the housing, fences, windbreaks, and gardens. This phase includes many different aspects that all need to be done at around the same time, which may seem overwhelming. Get your house up first, or on a property that has an existing house, work on modifying it to suit your needs. To establish the gardens quickly, put together a plant nursery to start the thousands of plants you will need for transplanting later. You can do this just before doing your house so that you will have a head start.

Phase 3—The last phase is a time of refinement. Implement your fire and erosion control, begin repairing damaged soil, and build more energy production systems. If you think you may want to put in a wind turbine at a later time, make sure that you planned it out in advance and preserved the space for it.

Self-Regulation

Finally, the permaculture process is a very long-term investment. Each piece of land is unique, and each design will be different. While the design process may seem a little complicated, it's just not possible to "do permaculture" wrong. Almost anything can be moved around to where you want it later. If some plants aren't doing very well in one spot, move them. Try another species and see if it yields more. The point is to observe, experiment, and learn from the natural world. The goal is to create a system that is *self-regulating*. That is, eventually you should reach a point when you can leave for a while, and when you come back, everything is still thriving. No water, no weeding, no hand was needed to keep it all alive. It should become a self-contained ecosystem that cares for itself.

The Successful System

- High *biomass* production. Biomass is the volume of living organisms in an area.
- Lots of organic matter. Even if it is not alive, the organic matter in and on the soil should be tremendous.
- Living organisms provide minerals rather than rocks and rain.
- Retains minerals over time. Consumption and erosion have little impact.
- Allows fungus and bacteria to play a central role to the cycles of life.
- Has a majority of perennial plants, keeping replanting to a minimum.
- Has a huge variety of species and massive diversity.
- Has complex food chains rather than simple ones.
- Uses big organisms rather than many tiny ones: trees, cows, and large plants.
- Sees relative sameness over time. Very little or very gradual change.
- Uses species that are useful in as many ways as possible.

⌃ **Everything makes a circle.**

that inhabit it. However, some awareness is necessary. The water cycle is the most obvious, with the wind and rain reminding us frequently of its processes. The soil goes through cycles that are evidenced by

NATURAL CYCLES

You will die but the carbon will not; its career does not end with you. It will return to the soil, and there a plant may take it up again in time, sending it once more on a cycle of plant and animal life.

~ Jacob Bronowski

Working with Cycles

Many of the cycles of nature will continue around you without your notice, even as you become more attuned to your gardens and the various creatures

⌃ **Chicken of the woods is an edible fungus and plays an active role in the garden.**

The Ultimate Guide to Permaculture

the plants that grow, the pioneer species fixing the nitrogen in the soil and preparing the way for other plants. Then there is the nitrogen cycle itself, which is just as crucial as the water cycle for life on earth. The atmosphere is ripe with nitrogen, but plants are unable to use most of it. Nitrogen must be chemically altered by bacteria (like *rhizobium*) or by lightning. Once it changes, it is *fixed*. The volume of fixed nitrogen in the soil decides how much will grow, and when those plants die, they release the nitrogen. There are also cycles of harmony between species. The ducks forage in the gardens in the fall after the harvest, which fertilizes the soil, removes pests and weeds, and feeds the ducks. They live and grow, providing eggs and meat for people.

Spring

- Remove the ducks from the gardens and put them in the marsh.
- Remove the geese from Zone 2.
- Prune the orchard trees.
- Check the rice paddy for thin spots and sow more seed if necessary.
- Remove mulch from Zone 1 gardens to allow soil to warm up.
- Check on the bees and make sure they have enough food and pollen.
- Watch for baby animal births.
- Shear sheep.
- Plant your earliest garden and harvest the last of the winter garden.
- Tap maple trees for syrup.

Late Spring

- Harvest rye or barley.
- Check on the bees to see if they have enough space and split them if you need to.
- Harvest your early garden.

Summer

- Trim back more orchard trees as needed.
- Flood the rice paddy.
- Plant more garden transplants.
- Harvest the summer garden and plant more transplants.

Early Fall

- Let ducks into the Zone 1 gardens.
- Let geese into the Zones 2 and 3 fields.
- Sow rice seed in the rice paddy.
- Harvest the last gardens and plant fall transplants.

Late Fall

- Harvest last year's rice and lay it out to dry.
- One month after rice harvest, plant next year's rice crop.
- Harvest your honey and possibly start feeding bees.
- Breed sheep.
- Harvest the fall garden.

Winter

- Check on the bees and make sure they have enough food.
- Dry up pregnant cows.
- Order new bees and seeds.
- Watch the ewes for lambing.

2 | Energy

The energy of the mind is the essence of life.

~ Aristotle

PASSIVE ENERGY

All peoples everywhere should have free energy sources.

~ *Nikola Tesla*

What it is . . .

Passive energy works in most places most of the time, and there are two kinds: solar and mass. There are also two kinds of mass: biomass and thermal mass. Often both types of passive energy work together to heat or cool something. Passive solar is just a fancy name for when the rays of the sun are utilized without any special equipment. Thermal mass is when a non-organic material that has good insulating properties is used to regulate the temperature of something. This material is usually very thick, or has a lot of mass. Biomass is when a large amount of organic material is piled up to create heat through the natural bacterial action of decomposition. All of these methods are low-cost, no-energy solutions.

The Passive Solar Home

The solar home must be very well insulated, with an open floor plan so air can circulate. All of the permaculture principles of house and window placement apply. The long side of the house should face north in the southern hemisphere and south in the northern hemisphere. The main living area should be on that side as well, with big floor-to-ceiling double-glazed windows. Passive solar can be as simple as allowing the sun to come in and warm up a room. Cover the windows at night during cold weather or during the day in hot weather to regulate the temperature.

Thermal mass is a large area of heat-absorbing material, usually stone or brick, which absorbs solar energy during the day and releases that heat at night. Wood, carpet, and furniture don't do this very efficiently. Once that solar energy enters your living room through the big double-glazed windows, it should fall upon thermal mass. You can create this mass by building the far wall or living room floor out of stone, brick, or tile. Depending on the layout of the house, you could build the wall directly behind the windows with small openings that let in light. The wall will radiate the heat into the room behind it.

Solar Hot Water

Solar hot water heating is one of the most efficient ways of harvesting energy from the sun. You can either purchase a commercially made system or build one yourself. The price to buy one has dropped dramatically in the last few years, making these systems a common sight in many places and also making

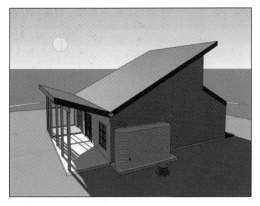

▲ Passive solar house at summer solstice.

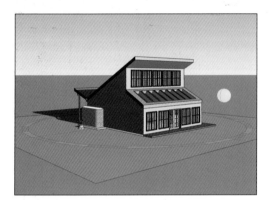

▲ Passive solar house at winter solstice.

it less worthwhile to build your own. There are five major types of hot water systems: batch, thermosyphon, open-loop direct, pressurized glycol, and closed-loop drainback. Each has its own advantages and disadvantages. In a cold climate, some are susceptible to freezing. Some are very heavy and thus may not be able to

be installed on your roof. Some are very difficult to install and maintain. Some need a pump and controls to operate, and some need an extra storage tank.

The most popular option is the thermosyphon system. Unlike the batch system, which heats a water tank directly from the sun, the thermosyphon system has a flat panel solar collector with piping going through it. This piping leads to a storage tank directly behind it. Based on the principle that water rises as it warms up, the cold water enters the pipes and, as it warms, moves towards the top of the collector and into the storage tank where it flows using gravity to the backup water heater. This is also a system that you can build yourself, although the parts and labor may be the same as buying one.

In a cold climate, the glycol or closed-loop system is used. These both use a liquid that won't freeze as easily and circulate it through the flat panel collector and then through a water tank inside a closed loop of piping. This is really just a basic heat exchanger. The glycol system uses antifreeze, and the closed-loop system uses distilled water. Once heated in a tank, the fluid then flows to the backup water heater. Both of these need pumps to operate, but they can be used in extremely cold climates.

Water Heater Type	No freezing	Light weight	Easy install	No pump	No storage
Batch			X	X	X
Thermosyphon			X	X	X
Open-loop		X	X		
Glycol	X	X			
Closed-loop	X	X			

The Ultimate Guide to Permaculture

▲ Solar design requires large windows on one side.

Before you can install a solar water heater, be sure to fix any drips and leaks in your faucets and install low-flow heads. Turn down the hot water tank to around 125°F (52°C) or lower and wrap it with insulation. Insulate your water pipes as well. The average American uses around 15–30 gallons (57–113 liters) of hot water per day, but this is entirely dependent on your own wastefulness. Run dishwashers and washing machines with full loads only, and take shorter showers. Your hot water tank should also have the capacity to hold enough hot water for your family for one day, at a rate of at least 20 gallons (75 liters) per person.

In a sunny southern area, you will need a solar collector that is one square foot for every 2 gallons (8 liters) of daily household use. For a family of four this would mean 40 square feet (4 square meters). In a cold northern area, the collector must be 1 square foot (0.10 square meters) for every 0.75 gallons (3 liters).

Don't expect to be completely free of electricity or gas to heat your water. It is true that during the summer you will probably get almost all of your hot water from the sun, but in the winter don't expect more than 40%. That means, however, that you are using your old hot water tank during only about a quarter of the year for only half the time, which is an eighth of the energy you used before.

What is biomass?

Biomass is organic material that is piled up to create heat. The act of piling it up isn't what makes the heat, but rather the action of the bacteria breaking down the material. You can use biomass passively through a *heat exchanger*, a device that transfers heat from one material to another, or you could use it to produce *biogas*. Biogas is the gas that is released during the breakdown of the organic materials. Most of the time it is manure that is used to make biogas in the form of methane. It takes about 240 pounds of fresh manure per day to produce enough gas to fuel a stove, and since a single cow produces 100 pounds of manure a day, it isn't hard to supply enough. However, we need to consider the principles of permaculture. First, the manure has to be collected when fresh, which requires human effort. This also means the animals must be confined. Since your animals would normally be foraging in edible forests and pastures while simultaneously fertilizing them, you would be taking nitrogen from your gardens and breaking the nitrogen circle. Biogas is also hazardous: When you mix methane and oxygen, they can explode, and so systems have to be foolproof. In a city or large community where composting isn't possible for many people, a large biogas plant makes sense. On a small piece of sustainable property, it does not, mostly because the amount of human manure and food waste does not supply enough gas to do anything.

Biomass Hot Water

So, biomass can be useful to us passively, but most of the supply of material will come from human food waste and composting toilets. Since that has to be broken down before going back into the soil, we can take advantage of the heat it produces. Implementing a small biomass system at home is as simple as placing the compost heap inside a building that you want to heat and running plastic piping full of water through it. The compost reaches a temperature of 140°F (60°C), and depending on the size of the pile, it can stay that way for six months to a year. The plastic piping can be a thin-walled PVC

▼ Water hose under the floor of a greenhouse built with bales of hay.

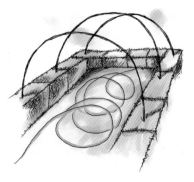

Climate	1 sq. ft. per daily gallons	Family of four (80 gallons)
South and sunny	2 gallons	40 square feet
Temperate region	1–1.5 gallons	80–120 square feet
Rainy and cold north	0.75 gallons	140 square feet

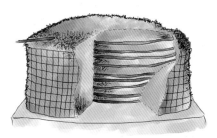

▲ **Biomass compost water heater.**

pipe, but you may have more success with a garden hose because it is even thinner.

The most common place to take advantage of biomass is in the greenhouse, where it will heat the room. But you can also do something better. Rather than using a wood compost bin, a ring made of chicken wire or other mesh fencing is set up to contain the pile. This circle is piled with a layer of compost and organic material until it is full. Then, a coil of plastic pipe is wrapped around and around it and another ring of wire is set up at least four feet (one meter) outside the first. More compost is added to that ring until it is full, and as it heats up it will heat any water that is run through the piping. This water can then be sent to the hot water tank for use in the house.

PHOTOVOLTAICS

I'd put my money on the Sun and Solar Energy, what a source of Power! I hope we don't have to wait until oil and coal run out, before we tackle that.

~ Thomas Edison

Parts of a Solar System

Solar array: A solar array is a series of glass sheets enclosing either single-crystal or poly-crystal solar cells on top of a waterproof backing material and edged with a mounting frame. Several solar arrays make a photovoltaic array.

Charge controller: A device that prevents overcharging the batteries during the day and discharging during the night. A cheap charge controller is simply a relay device that opens and closes the circuit to the battery. Invest in a better controller that will keep the batteries above 20% and make sure the voltage isn't too high.

Battery bank: One or more batteries wired together matching the voltage rating of the inverter and the solar array. These must be deep-cycle batteries, which can handle the charge/discharge cycle. Car batteries will not work for this.

Inverter: If you are using 120 volt AC lighting and appliances, you will need an AC inverter. The inverter is a crucial component and has to be able to handle the load you need it for. A marine or RV inverter won't work for a house. Well pumps, washing machines, and refrigerators need a large capacity sine wave inverter. Those cannot operate from a 12-volt battery bank and so you would need a solar array and battery bank designed to operate at 24-28 volt DC.

Safety fuses and circuit breakers: Fuses and circuit breakers are necessary safety devices that must be DC rated.

Installing a Photovoltaic System

A solar array can be mounted on a tilted rack or on a pole or put on the roof of your house or garage. The angle depends on your latitude. In the northern hemisphere tilt your panels towards the south, and in the southern hemisphere tilt them towards the north. A steep tilt in winter and a shallow tilt in summer will improve performance.

The battery bank must be big enough to store energy for at least three or four days, which means you might need quite a few units. These must be stored in a battery box and should not be exposed to freezing or very hot temperatures.

When the sun shines into the array, the array converts the sun energy into electricity. That power is sent to the regulator, and the regulator sends it to the batteries and inverter. The inverter converts DC to AC and sends it to the AC electrical equipment. A very frugal household usually uses 1,000 kWh per month, or 34 kW per day. In most places you can collect five and a half hours of sun in summer and four hours in winter, which means that if you didn't improve your energy-using habits, you would need a 6-7 kW array. A system that size can cost $25,000 (US) or more. If the furnace, dryer, hot water heater, and stove are not hooked to it, a 4,000-watt inverter will take care of a house. A photovoltaic system will probably only harvest 65-75% of the advertised rate, so that has to be planned for as well.

Even by taking those power-saving steps, a system completely reliant on solar energy would still cost upwards of $15,000 (US).

The wiring of a solar electric system must be built to code and has to be inspected by a local building inspector. Because of this, unless you are an electrician, it is probably a good idea to hire someone to install it. The price is high, and so a solar electric system is usually used in conjunction with a solar hot water heater, wind power, and many energy-saving devices and habits.

Maintenance

Maintenance on a solar system is much less than on other systems but still requires some routine checks on the batteries. Your batteries are your greatest expense, so keeping them in good shape can save you a lot of money. Be cautious when messing with the batteries as they can release toxic gases or explode.

- Regularly check the water level and fill it with distilled water only.
- Keep the battery tops and terminals clean.
- Always keep the batteries charged, even when you're not using them.
- Protect them from freezing.
- Get as many batteries as you can in the beginning to decrease the load on each one.

▾ **Solar array.**

Latitude	Angle of array
0–15°	15°
15–25°	15–25° (same as latitude)
25–30°	30–35° (add 5°)
30–35°	40–45° (add 10°)
35–40°	50–55° (add 15°)
40° +	60° + (add 20°)

The Ultimate Guide to Permaculture

- Use a good charge controller that makes sure the charge never drops below 20% or higher than 20%.

Panels mounted on a stand are usually maintenance free, but panels on a roof require regular cleaning and protection from snow. Heavy snow can crush the panel, and leaves and dirt can get trapped and decrease the panel's efficiency. It is a good idea to keep a supply of spare fuses and breakers for the DC system in case something breaks. The controller can also greatly prolong the life of your batteries, which are the greatest expense in a solar system, so invest in a good one that will regulate them well.

Viability of Photovoltaics

While every climate is different and it is impossible to predict how much sun you will get, solar power always makes the most sense in the desert. At the time of this writing, you cannot build or install these systems yourself without extreme cost, and so unless you get at least 150 fully sunny days a year, it is probably not worth it. In some places government subsidies exist to help offset the cost of purchasing a solar power system.

While it is important to be off the grid and use clean power, this only works if the household decreases its power consumption dramatically, which means a drastic lifestyle change. The system should be able to pay for itself in savings, or else it doesn't make sense. A small system can run the lights and help run the backup hot water tank in the winter when the passive solar one can't work, a very small fridge, and perhaps a low-energy laptop computer. That much is probably worth it. If you have a well with an electric water pump, build a cistern so you can use a pump with lower wattage. This sized system would cost about $3,000–$5,000. The stove, water pump, televisions, and other devices would need to have a wind turbine or microhydro to run. Solar power research is progressing very quickly, and the price should drop dramatically in the future.

HYDRO

When you put your hand in a flowing stream, you touch the last that has gone before and the first of what is still to come.

~ *Leonardo da Vinci*

The Water Source

To produce electricity from water, the first thing you need is a flowing stream or river that runs year round. That may not even be enough to produce electricity, however. To learn if it can, you must find out its elevation and how fast it is going in feet per second. Once you have verified those numbers, you can buy a small hydroelectric generator, which is about the same price as a small diesel generator.

In the old days waterpower generation was produced by means of a water wheel, and even today in large hydroelectric dams, turbines are a modification of these wheels. However, for the small or *microhydro* setup, you can purchase small turbines or pumps. This choice is dictated by the flow of your water.

Pelton wheel: The Pelton is a small high-speed wheel that pressurizes the water through a spout so that it hits the wheel with more force. This wheel is popular because it has an efficiency rate of 70–90% and can work in streams that are slow.

Turgo wheel: The Turgo is an improved version of the Pelton that angles the nozzle to hit more of the paddles of the wheel at once.

Jack Rabbit wheel: The Jack Rabbit looks much like a small boat propeller and is simply dropped into the water. It can run in a very shallow stream with very little flow. It produces less power, around 1.5–2.4 kWh per day, which could power a *very* frugal household.

Pumps: A hydraulic ram pump can be used to pump drinking water to the house without electricity, but if you want to use a pump to produce power, then the water can be run through it. Normally a water pump is used to move water, but you can use water to move the pump. It is less expensive to get an electric pump than a water wheel.

It is difficult to regulate the voltage produced by a water turbine, so using AC can be practically impossible. DC allows you to store electricity in batteries and even use car parts to build it. A car alternator can be hooked to a wheel or pump to create the electricity. If you have solar, wind, or diesel systems, DC also lets you link them together.

Generally, a stream has *low head* (head is the amount of water pressure) if it has a change in elevation of less than ten feet, while a stream with *high head* has more than ten feet. If it has a drop of less than two feet, then you will have to use a Jack Rabbit as long as the stream is deeper than one foot. A mathematical formula is used to find out the head of a stream.

To find out your net head:

First, find your total gross head, which is just the stream's drop in elevation. If you

▲ Hydraulic ram pump.

have an altimeter (some watches or smart phones have them), you can use it to figure this out easily. Measure the drop from the water source to where it will come out of the turbine. One low-tech way to do this is to get a 20-30-foot (6-9-meter) hose, a funnel, an assistant, and a measuring tape. Have your assistant stick the funnel in the end of the hose and put it right under the surface of the water at the source. Hold the other end downstream and lift it until the water stops coming out of it. Measure the distance between the water and how high you had to lift the hose, and that is your gross head.

Now you need to find your minimum water flow. Dam the stream with logs or boards to divert it to a place where you can place a 5-gallon (20-liter) bucket. When the stream is flowing into the bucket, use a watch to calculate how many minutes it takes to fill. If it fills in two minutes, then your flow rate is 2.5 gallons (10 liters) per minute. Repeat this several times a year because your stream's flow rate will fluctuate depending on the season. The minimum water flow is the lowest that it gets during the year.

Calculating Kilowatts

The formula for calculating power is this:

The Ultimate Guide to Permaculture

(Gross Head)x(Flow)x(System Efficiency decimal equivalent)x(C) = Power (kW)

- C is a constant and equals 0.085.
- Gross Head was the elevation of the stream.
- Flow, which we had calculated in gallons, needs to be converted to cubic feet per second. To convert gallons to cubic feet per second, multiply by 0.8327.
- System efficiency is 0.55 because a well-designed hydroelectric system needs to run at an average of 55%.

Example:

(15 feet) x (2.5 gallons = 2 cubic feet per second) x (0.55) x (0.085) = 1.4 kWh

If we compensate for the loss of efficiency because the wheel itself slows down the water a minimum of 30%, we end up with 1 kWh. This means that over a day we can produce 24 kW, and over a month we can produce 720 kW. Water is one of the most viable power options because it produces power 24 hours a day. With a battery bank, it would be possible to make up the difference in power needed. As we learned from the previous section, a frugal family needs 1,000 kWh per month.

WIND

First, there is the power of the Wind, constantly exerted over the globe . . . . Here is an almost incalculable power at our disposal, yet how trifling the use we make of it! It only serves to turn a few mills, blow a few vessels across the ocean, and a few trivial ends besides. What a poor compliment do we pay to our indefatigable and energetic servant!

~ Henry David Thoreau

Viability

A wind generator will power your entire house if you have at least 8 mph winds most of the time. The cost is also relatively inexpensive. Wind generators usually cost more than hydroelectric but less than photovoltaic ones, so if you live in a windy area, then it may be the best option. Part of the reason they are less expensive than hydro generators is that they are usually less efficient. A wind turbine has three rotor blades and a tail to keep it aligned with the wind. The diameter of the rotor blade wheel determines its *swept area*, or how much wind the turbine intercepts, and this can be applied to a formula to find out how much power a wind turbine can *theoretically* produce. It's a good idea to do this yourself rather than rely on the manufacturer's numbers, which are often extremely optimistic.

0.5 x (Wind Air Density) x (Swept Area in Meters) x (Velocity)^3 = Power (W)

- Wind Area Density is a constant = 1.23.
- Swept Area is the area of the rotor in meters squared = π x radius^2.
- Velocity is the speed of the wind, cubed, in miles per hour.

Example with 8 mph winds and a rotor with a diameter of 2.4 meters:

0.5 x (1.23) x (3.14 x 1.2m^2) x (8 x 8 x 8) = Power (W)

0.5 x (1.23) x (7.536) x (512) = 2,370 Watts

Unfortunately, your calculations aren't complete yet. The turbine itself slows the wind, and the best you can possibly get is 59.26% of its capacity. For the above example, the most power you could generate is 1,420 Watts. Add to that the imperfections of the blades and the wear

▲ Small home wind turbine.

and tear on the rotor, and efficiency can drop to 30%. Any turbine that you buy that claims efficiency above 59% is misleadingly labeled, and it is unrealistic to expect more than 30%. The real power output of the above example would then be 710 watts.

As we learned in the previous sections, a frugal family uses about 1,000 kWh per month. At the rates of the previous example, you could produce about 17,000 watts, or 17kWh (kilowatt hours) per day, or 510kWh per month. You would need a bigger rotor or a higher tower, or you will have to use two turbines.

The height of the tower can greatly increase your power output. It is a waste to build a short tower, and you can't install it on your roof, not only because it won't generate much power but also because the steady vibration of the rotor will wear your roof out. The rotor blades should be at least thirty feet above anything within 300 feet of the tower. The higher you go, the more wind you will have, and you are also more likely to avoid turbulence, which can also decrease your efficiency.

There are three main tower options:

Freestanding: A freestanding tower is the old-fashioned style that you often see attached to windmill pumps. It has metal framework with four legs that are supported by each other. They take up the least space but are often the most expensive. These are the easiest to climb for maintenance.

Guyed: A guyed tower is a long pole that is help up by three or four guy wires. They are the least expensive but may be the trickiest to install as you often have to either have a group of people or a crane.

They are also difficult and dangerous to climb for maintenance.

Tilt-up: Tilt-up towers are similar to guyed towers, but their guy wires are designed in such a way that the tower can be winched to the ground for easy maintenance.

The power from the turbine is directed to an inverter, just like the power from a solar array or hydro turbine. A wind turbine needs regular maintenance, but in general it only needs a checkup every six months. You can also use a battery bank to help store up power for windless days.

Installation and Maintenance

Wind generators are fairly straightforward. You buy one, anchor it in a windy spot, and connect it to your system just like your solar panels. The top of a hill is usually a good place, away from trees that may eventually grow to block its way. Be mindful of your neighbors and local zoning, which may not appreciate your apparatus or may have restrictive bylaws. Some turbines are quite tall and, depending on their size, can create quite a bit of noise.

Noise is also a factor when purchasing a turbine. A silent turbine doesn't exist, and any that claim to be are full of it. Some of the untrue made by wind power companies are: They don't kill birds, they work in barely any wind, or the turbine is a new, patented design. Many of the people selling turbines don't know anything about wind power. This doesn't seem to happen so often with other forms of home power generation. Be cautious and remember that the formula above works no matter what the design of the turbine is. The only factors that increase your power output are the diameter of the rotor and the height of the tower.

Once installed, the turbine will need to be oiled and checked every year for any loose bolts, wires, or screws. Some turbines are made of wood and need to be repainted for protection. In the winter, you may have to de-ice the blades. A wind turbine also has a *survival speed*, or a maximum wind speed that will make it lock up, and in especially strong winds it could sustain damage. However, a properly maintained turbine can last 20 years or more if the blades and bearings are replaced when necessary.

3 | Water

I understood when I was just a child that without water, everything dies. I didn't understand until much later that no one "owns" water. It might rise on your property, but it just passes through. You can use it, and abuse it, but it is not yours to own. It is part of the global commons, not "property" but part of our life support system.

~ *Marq de Villiers*

WATER SOURCES

All streams flow into the sea, yet the sea is never full. To the place the streams come from, there they return again.

~ Ecclesiastes 1:7

Assessing the water situation:

Water is the most important element. On a piece of land, no other factor impacts so many other things or is affected by so many other elements. How much water you have depends on the rainfall, how the soil drains, the plants that are currently growing, the people and animals using the water, and the kinds of plants you want to grow. Some of these factors are in your control. The first step is to decide where the water is coming from and devise a way to store it, utilizing gravity to move it to where you want to use it. The second step is to use species that need less water in places to which it is difficult to get water.

While trying to reach these goals, keep in mind that water is also a responsibility. Your job is to use the water you harvest for as many tasks as you can and be super efficient when you do: growing life through gardens and aquaculture, and generating electricity. Growing life and generating power can be done together and be mutually beneficial. You must also conserve and reuse water. Use as little as possible and recycle it as many times as you can.

Sources of water:

Rain—surface runoff or groundwater.
Springs—groundwater.
Stream—permanent or temporary streams.

These sources are *diverted* through various means to the places that we store and ultimately use the water. We do this by taking advantage of the free and energy-efficient power of gravity. Gentle slopes and drains lead water from streams, valleys, roofs, and roads that collect rainfall and send it to storage, irrigation systems, and *swales*.

The desert provides us with warm temperatures all year round with one drawback: very little water. Rainfall is usually brief and produces less than an inch of rain, which runs off the sand soil. This runoff can be diverted with drains made of dirt, concrete, rock, or pipes to a tank or into a basin or terrace. The basin should be designed to handle runoff from land 20 times its size: 20 acres of land would feed into one acre of native trees or crops. When we do this, it is important to plan ahead and create channels that are resistant to erosion. The desert also usually has an interesting landscape because of the dry sandiness of the soil. During brief torrential rainstorms, water runs off the land into valleys and becomes a raging

▾ Topography and water flow.

river within minutes. These flash floods erode the soil quickly, which has shaped the desert into its characteristic mesas, canyons, and flood plains.

The soil quality dictates how much runoff you will have. A sand dune will absorb all of the rainfall, and a treed or grassy meadow will absorb a little less. Sandy soil will absorb less than a sand dune, and undrained clay will absorb almost none at all.

SWALES

Water is the driver of Nature.

~ Leonardo da Vinci

What is a swale?

A swale is a long, shallow ditch about three to five feet across, which serves to stop and channel the flow of water into

▴ A swale right after a heavy rainfall.

the soil. Unlike a regular ditch, which directs and carries water somewhere else, the swale is made to help the water get absorbed directly into the soil. Swales lie across the land, especially across a slope, and when the water is forced into the soil, it can be soaked up by trees planted along the swales. Swales can be filled with rock, gravel, or gypsum for even better water penetration. Swales are perhaps the most effective method of water conservation in both dry and humid climates, they work well on steep slopes or on flat prairies, and they can even be implemented in an urban area to take advantage of road and roof runoff.

Building a Swale

It is not difficult to dig a ditch a few feet deep. The swale will catch the water in a pool as the rain runs into it, where it will gradually soak in. If the swale is overflowing, then you need to widen it or improve the drainage. Over time you can throw mulch in, and grass will probably grow. As the trees get taller and begin to shade it, other species will spring up.

The swale is really all about positioning. Two or more are always put in together, along the counter of the slope, and the soil is loosened up to help water penetrate. The second swale is placed 10–60 feet away from the first, depending on the amount of rainfall you get per year.

▴ Large swale curved to follow the land.

The Ultimate Guide to Permaculture

Rainfall	Space between swales
5 inches	60 feet
10 inches	50 feet
20 inches	40 feet
30 inches	30 feet
40 inches	20 feet
50 inches	10 feet

After a couple of rainfalls that soak deeply into the soil, plant trees directly behind the hump of dirt. In a wet climate the area between the swales can be heavily planted, and in a dry climate only the edges of the swales should be planted. It can take three to ten years for the trees to fully mature and begin their job of shading, mulching, and absorbing salts from the soil. This is the most important step of the whole swale strategy. Trees will soak up the water runoff and prevent waterlogging and evaporation produce valuable products and food, and improve the temperature and microclimate of the land.

DIVERSION

How beautiful is the rain!
After the dust and the heat,
In the broad and fiery street,
In the narrow lane,
How beautiful is the rain!

~ Henry Wadsworth Longfellow

Drains

Unlike a swale, which stops the water and forces it into the soil, a *diversion drain* is a ditch that carries water away. A diversion drain can be used to direct water into a swale or to a pond or irrigation system. Drains are also used to direct the flow of water to the series of dams you would build to catch water runoff. If they are directed to swales,

they don't need to be waterproof, but if you are sending the water to a dam, the drains should be built of rock or concrete. Installing a *spill gate* also gives you control over the flow of water. If you fall victim to flooding or you want to control crop irrigation, you can use the spill gate to direct the water where you want. A spill gate is simply a removable method of blocking water.

Dry Bed Management

This kind of management is most valuable in the desert where small creeks and streams can quickly get out of hand after a short rainfall. The rain runs into the low creek beds and rapidly becomes a flood, eroding the soil and becoming unusable to anyone. It flows too fast for it to absorb into the soil or to be taken up by plants that conglomerate in the creek bed. It may also cause damage.

Stream *braiding* is a non-technical term for spreading the flow of a stream out into a myriad of much smaller channels over a landscape, so that it irrigates the entire area and at the same time prevent flooding. This happens naturally in many areas where silt and sand erode and deposit downstream, breaking up the terrain. To do this, start at the head of the

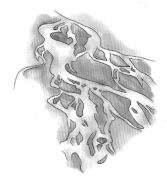

▲ **Natural braided stream.**

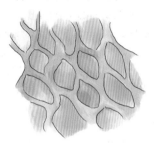

▲ Man-made braided stream.

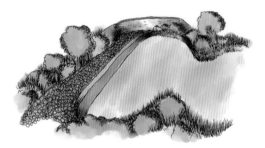

▲ Flood dam slows water flow and directs it to a reservoir.

stream and dig a small pool. Branching off from this pool, you will create a diamond pattern across the landscape by digging diversion channels. Where each diamond intersects, you will dig another small pool. This strategy works well for streams that tend to run dry parts of the year and are prone to flash floods.

The small pool strategy can also be used to manage the flow of a stream that intersects an area that can't be spread out, such as near the house or barn. Instead, a pool is dug at the head of the stream to create a lagoon, while swales and dry dams are installed between the lagoon and the building.

In a gully or floodplain where streams flood over a wide area, you can use a dam to trap water, but not all at once. The dam will be less than 20 feet high and short enough for some of the floodwaters to spill over. Instead, the dam is wide, slowing the water down as it climbs over the wall. Behind the dam, the direction that the floodwater comes from, the sides of the gully can be carved out into curves to disrupt and slow the flow of water. Then plant trees and grass on both sides of the dam to slow things down even more.

Once you've slowed the flood, you can divert some of it to the side. The dam extends longer than necessary to the right or left and slopes into a reservoir with a

tall embankment to keep the water in. The wide top of the dam, while not stopping water, acts as a channel to the reservoir. In the desert you will not be able to keep the water in the reservoir, because it will evaporate. It must be channeled to irrigation or a tank right away.

Please note that this is for extreme flood situations only. This kind of dam is not meant to block a major waterway but rather is put in a dry bed that experiences flash floods. The reservoir is put in place to try to trap some of that excess water.

Rainwater Catchment

Every 1,000 square feet (92 square meters) of roof surface area will gather 600 gallons (2,270 liters) of water per inch (2.5 cm) of rain, and every roof on your property can be used to collect rainwater. The best material for this purpose is metal because it's relatively clean. You will also need to install gutters with a leaf screen.

If the rainwater is being used for human consumption, it should be directed into a roof washer that automatically diverts 1 gallon (4 liters) per 100 square feet (9 square meters) into a separate tank or to the garden. This will help keep dirt out of the water. The entrance of the stored water should be covered with a mosquito screen, and the tank should also have an overflow valve directed into the garden.

▲ **One type of roof washer, using a tube.**

1. From the tank the water will go through a series of pipes to where you want it to go. On the house, it is necessary to set up some check valves to prevent any back flow. Unless your tank is far enough up a slope to provide gravity pressure, you will also need a water pump.

2. From the water pump, there should be another check valve and then a pressure tank. There are other options, such a water-sensitive pump, which will create water pressure *and* pump your water in one device, but the most common solution is a small tank, which pressurizes water for immediate use. This tank must be indoors in cold areas to protect it from freezing.

3. Once in the house, the water would then go to a water purification system. After that it can finally enter your pipes. If you are in the city and still want to be connected to the municipal water supply, you usually need an approved backflow prevention device that will ensure no rainwater gets into the city water supply. This device must be examined and maintained properly.

TANKS

We can't help being thirsty, moving toward the voice of water.

Milk drinkers draw close to the mother. Muslims, Christians, Jews, Buddhists, Hindus, shamans, everyone hears the intelligent sound and moves with thirst to meet it.

~ Jalaluddin Rumi

Human Water Consumption

Water for human consumption is stored in tanks. A single person should store at least 1,000 gallons (3,700 liters) of water, and a family needs a minimum of 2,500 gallons (9,500 liters). Tanks can be made from concrete, compacted clay, metal, or even plastered dirt. The water to fill the tanks comes from rainwater that runs off roofs of buildings or other surfaces or is pumped from a dam. To repel mosquitoes, the tank should be covered and screened. Thick green algae will begin to grow over the sides of the tank, but this is a good thing because the algae will help clean the water. The outlet pipe of the tank can be 3 inches (7.6 cm) above the bottom so that the algae remain undisturbed.

Some people have recommended that mosquitofish can be used in a tank to eat mosquito larvae. The use of mosquitofish (or gambusia) does more damage than good, and it is questionable if they eat any more larvae than any other fish. They have hundreds of fry and breed prolifically, but they only live a couple of years. They should not be introduced to your aquaculture pond, because other fish tend to avoid eating them and they can quickly choke out other populations. A more effective mosquito control is frogs and birds.

It makes sense to place a dam or water tanks at the top of a hill. In fact, a large water tank set on the top of a hill can act as the foundation for a building, and the building roof can be used to collect rainwater. The house will have its own water tank, and the pond or lake can be situated below the house at the bottom of the hill, acting as a firewall and reservoir for drought.

Choosing a Tank

There are many different types of water tanks. Choosing the right type can be difficult because the pros vs. cons aren't always sufficiently clear. Price is also a factor. The earthbag tank is cheap and durable but takes up a huge amount of space. Polyethylene is cheap and durable and comes in any size and shape, but people concerned about chemicals leaching from the plastic will want to avoid them. Even an earthbag tank is lined with polyethylene, and so it may not be any cleaner. Galvanized steel is considered safe to drink from, although it has a zinc coating that does leach into the water. It is a good thing that the zinc no longer has lead in it, which manufacturers at one time used. The concrete tank is less expensive, but it has more metal in it in the form of rebar than

▲ Gravity fed water from dams and rainwater collection at the top of a slope.

a galvanized steel tank, and even then its lifespan is shorter than that of other options. However, if the right concrete is used, it also may be the safest to drink from.

Earthbag Cistern

Earthbags make a very inexpensive water tank. You must use 50 pound bags made of polypropylene or other durable plastic that will be waterproof and at least 17 inches wide and 30 inches long. See the section on earthbag houses for more information.

Tank Material	Pros	Cons
Galvanized steel	Durable Takes up little space Lasts 20 years	Must have a concrete base Can't be set in the ground More expensive Possible metal leeching
Polyethylene	Available in all shapes/sizes Less expensive Doesn't need a concrete base Durable Lasts 25 years Takes up little space	Possible chemical leeching from plastic
Concrete	Less expensive No chemical leeching	Lasts 15 years Has more metal in it than a steel tank (rebar) Takes up more space
Earthbag	May be cheapest Lasts 20+ years	Takes up more space

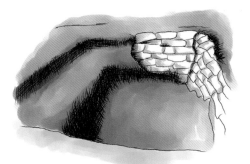

1. The tank is set at least a quarter of the way into the ground, so you have to dig a smooth hole the size and shape of your tank, leaving room for the earthbags.
2. Line the hole with pond liner or heavy polyethylene.
3. Begin building the walls. As you put down each layer of earthbags, lay two strands of barbed wire between each layer to hold it together.
4. Once you have the walls done, they need to be waterproofed. You can plaster walls with concrete or stucco, lay down a pond liner or use a sheet of heavy polyethylene. Cover the top with an earthbag dome or concrete.

Cold Climates

In a cold climate the water is likely to freeze over and make a certain amount of your water unusable. Several feet of ice can make a significant difference in your water supply. Pipes from tanks to the house must also be buried in the ground at least three feet to prevent freezing and bursting, or they will have to be left on all winter. Such wastefulness can also quickly drain your tank. You can prevent many of these issues by putting the water tank under the ground, possibly under the barn, uphill from the house.

DAM

**The river seeking for the sea
Confronts the dam and precipice,
Yet knows it cannot fail or miss;
You will be what you will to be!**

~ *Ella Wheeler Wilcox*

The Strategy

A dam serves two purposes: watering animals and storing water in case of drought. Dams don't work well in dry climates. When water is stored openly in a hot, dry place, it just evaporates, increasing the salt concentration of the water. Dams are much more useful in a wet or humid climate. The dams described here are built in a series together, moving down from the highest elevation to the lowest valley in order to catch and store as much water as possible. These are low barrier dams with walls no higher than 20 feet (6 meters), usually situated on a slope of 5% or less. This keeps construction costs down.

Saddle: These dams are built on the lowest point between the tops of two hills, or the *saddle*, to catch rainwater runoff. Being at the top of the slope means that water can run downhill without the need for pumping. The dam can be a hole dug in the ground, or walls can be built up to

▾ From top left to bottom right: saddle dam, horseshoe, keypoint, barrier, and contour dams.

form the barrier. This type of dam is used to water livestock and for storage.

Horseshoe: Named for its shape, the horseshoe dam is built on a slightly flatter ridge of a slope below the saddle dam. Several are built on descending ridges to catch overflow from the saddle dam and rainwater runoff from the slope. This type of dam is used to water livestock and for storage.

Keypoint: The keypoint dam is built in the valleys of small streams along the contours of the terrain. Because streams always run downhill, they are excavated at the highest point of whatever slope exists. These are used to store water for irrigation.

Barrier: At the bottom of this series of dams, a larger barrier dam is carefully constructed across a streambed with a very large *spillway* that directs overflow into streams via the contour of the land.

Contour: These are built in the flattest areas and are excavated along the curve of the land. They are used for irrigation and raising fish and may also serve as a place to retain floodwater in desert areas that is directed there via diversion drains.

Building the Dam

No matter what the slope of the terrain, the series of dams built for storing rainwater runoff should never have walls higher than 20 feet (6 meters). You may be imagining a slick concrete wall protruding from the side of a hill, but that isn't the cheapest or the easiest option. The dam itself is more similar to a small pond, with a gentle slope on all slides. It is scooped out of the side of the hill, and the clay is piled out to make the wall. The outer slope of the dam should be roughly the same angle as the inner slope.

The dam also needs a *spillway*. The spillway is an outlet for water near the top of the reservoir that directs overflow where you want it to go. Rather than filling to the top and spilling over in a waterfall, the spillway would be able to direct the flow to the dams and ponds below. While you can plop a dam down in most places, the spillway location and maintenance is essential. Without it, water can quickly become a destructive force that erodes the soil or damages structures. The spillway should be grown with grasses and vegetation at least 8 inches (20 cm) high and kept free of debris.

It is unlikely that your soil will be waterproof enough to keep water in. Your dam will leak unless it is sealed. An old method of doing this is covering the dirt with manure or *gley*. Fresh manure is smeared in an 8-inch (20-cm) layer over the inside pool and the outer walls of the dam, then covered with another thick layer of dirt, cardboard, plastic sheeting, or some material which will promote fermentation. If you use sheeting, it needs to be weighted down with rocks. In a temperate climate it will take about a week for the manure to ferment into an airtight mass. Once it is done, fill it up with water and test it out. You can remove any plastic or carpet later.

One very simple method of construction that has already been talked about in this book is earthbags.

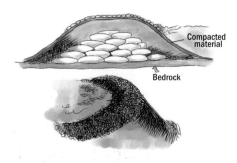

Compacted material

Bedrock

You can use earthbags in the same way that you would build a water tank, with polypropylene bags and covered with pond liner or polyethylene. You can even plaster it with concrete. This is a cheap, long-lasting solution.

In the desert the principles are the same, but the terrain is more extreme and the uses are different. The dams are placed at the top of plateaus, which are not completely level or flat, and divert water to storage tanks. While the walls of a mesa are usually very steep, they are still natural terraces and graduated slopes near the bottom that can be utilized for water collection with the same types of dams listed above. These divert water to the valley below where crops and animals are grown, and very little is actually stored for any length of time.

Desert dams will also still collect silt and must be cleaned now and then. It is doubtful that you could get anything to grow at the top of the plateau, where you would typically plant trees in a different climate, unless you built rock containers for them. The valley may also fall victim to flooding and since this is where you are growing food, it becomes necessary to try to stabilize the flow of water. *Dry dams*, or earth walls with rock spillways, can be interspersed with swales and rows of trees to slow down the flow of water and force it into the soil. Diversion drains are then carefully placed to flow runoff water into reservoir.

PONDS

A pool is the eye of the garden in whose candid depths is mirrored its advancing grace.

~ *Lousie Bebe Wilder*

Pond Design

Water holds and reflects heat. As the temperature drops at night, heat radiates into the air and surrounding gardens. A dam or pond will keep plants and people warmer in the cold, cool them off in the heat, and serve as a place to grow water-loving plants and fish. Tiny frogs will live there and eat bugs. Your pond can grow water chestnuts, rice, bait fish, brine shrimp, snails, aquarium fish, water lilies, basketry materials such as reeds and rushes, and mushrooms. It can also be a home to crayfish, prawns, mussels, clams, and ducks.

Some rules of thumb for designing ponds:

- For fish, several small ponds no more than 4 to 6 feet deep work better, while storage ponds can be 10 to 20 feet (3 to 6 meters) deep.

- A large pond for irrigation shouldn't have many fruit or nut trees around it, or the pond will become polluted with leaves and dropped fruit.
- Maximize the edge by creating a winding shore rather than a smooth oval or circle.
- To build a tiny pond in Zone 1, any waterproof container can be used. You can use stones to disguise the edges of a bathtub sunk into the ground.
- Large livestock can't have access to ponds for aquaculture, or they will destroy the balance of the pond.
- The pond must be near a water source to be filled initially or to top it during a dry season. The source can be a dam, diversion drain or hose.

The pond should have various refuge areas for each species that lives there. For the ducks, a small island can be formed in the center. A series of shallow shelves around the edges can serve as a home for forage plants. The pond also needs to have some deep areas at least 6 to 8 feet deep (2 to 2.5 meters), where fish can retreat in the summer when it is hot. Drop in some hollow logs or pieces of pipe for them to hide in.

The depth or volume of the pond doesn't have any impact on how many fish you can stock. The population of the pond relates to the total square feet of its surface. A pond 3 to 6 square feet (1 to 2 square meters) can be used to raise some plants and a small population of fish. A pond at least 150 square feet (14 square meters) can supply a family with all the fish they need (if the right species are selected), water plants, and a flock of ducks.

Pond design isn't extremely complicated. As long as you utilize any swampy areas or low-lying ground you already have and waterproof the bottom, you can build a fairly successful pond. Keep in mind the rules of thumb to be even more successful at growing things in it.

Constructing the Pond

1. Lay down a rope as a guide for the shape of your pond while you dig.
2. Divide the pond into thirds: one-third for the shallow end, one-third for the mid-range depth, and one-third for the deep end.
3. Line the bottom with pond liner or *gley*. Pond liner is a heavy black plastic, rubber or geotextile sheet made specifically for this purpose. The edges of any small pieces should be taped together or weighed down with heavy rocks. You may also want to put down a layer of pond underlayment under the liner to protect it. Gley is a layer of fermented manure. This is described in "Building the Dam" on page 64.
4. The edges of the pond will need to be stabilized with rocks of varying sizes, bamboo, grass ledges, or logs, especially at first, before other plants step in to prevent erosion. This step also helps hide the liner, so it's a good idea to use several materials for a natural appearance.

▼ Aquatic plants.

5. Do not introduce fish right away because there won't be enough food. Once you have built the dam and the pond begins to fill, lay down a couple of inches of straw and trample it into the bottom. This will provide a habitat for water insects.

6. Introduce water plants such as lilies, water chestnuts, and duckweed to build up the bug population. This process may take at least six months. The plants have to be very well established to withstand being eaten all the time by fish and other animals. You can start this process by leaving the plants in plastic pots for a while and simply submerging them in the water, so you can move them around if you need to. Transplant them when they get bigger.

7. The pond inlet from any outside water source must be planted with grass to help filter the income water. Keep the inlet clear of any debris.

8. The water should become green. If it doesn't, add a small amount of manure. The water may also be muddy at first if it is coming from a dam or other flowing source. If that is the case, add 1 teaspoon of gypsum per square foot (4.5 grams per 0.10 square meters), or 486 pounds per acre (220 kg per 0.4 hectares).

9. If the water gets too warm or it has been cloudy for a while, you may need to aerate it or the oxygen in the water may drop too low. Plants won't oxygenate the water unless it is sunny, and they may not be able to keep up the demand in hot weather. The average water temperature should be about ¾ the air temperature. This is why people have tiny waterfalls and pumps that keep the water flowing.

However, we are trying to save energy, so you might want to invest in a floating emergency aeration system. This is an electric device that sits on the surface of the water and creates bubbles and disturbance.

10. Hopefully, you won't need an emergency aeration device if you have the right proportion of plants. The plants should cover around 60% of the surface of the pond. Plants that live under the surface should be spread at a rate of one bunch per 2 square feet (0.2 square meters)

11. Introduce baby fish in the spring. You can also add a few buckets of pond water from a nearby pond to introduce a supply of aquatic insects.

12. Over time, the bottom of the pond will acidify, and while you can add lime to balance it out, every few years you may need to drain it. When it is dry, you could raise a crop of melons or rice. Your primary goal is to maintain water quality, with a pH of around 7 or 8.

Groups of Ponds

Several ponds situated in a succession of locations down a stream can be utilized to segregate fish of different ages. In this situation, several smaller ponds and marshes are connected to each pond to supply food to the main pond, and these

▾ **Keeping several ponds close to each other is useful.**

are all protected from predators. The only negative side of this setup is that any disease or parasite that invades a pond at the top will invade the lower ponds as well. A similar setup that solves this problem is to build the ponds parallel to each other rather than downstream.

Another alternative is a canal. Fish that eat plants, such as *tilapia*, are particularly suited to this kind of slow-moving flow of water with lots of vegetation along the sides. Canals also make it easier to catch fish with nets. This method can be used in conjunction with a garden, such as that of chinampa design (see aquaculture in the next section).

Aquaculture

Aquaculture isn't just a pond. It is a closed-loop system of growing fish in a cycle with plants and other animals. These range in size from small backyard ponds to large intensive aboveground tanks. Historically, the South American chinampa structure of canals stretching between large planting beds was filled with fish and was one of the earliest examples of highly productive aquaculture. Because aquaculture does have such high production value with relatively little effort, it is very tempting to set up the large tank system and turn it into intensive agriculture. Some very profitable methods have appeared in which one species of tank-grown fish are grown with plants in an aquaponic system, but this book deals chiefly with the natural pond version. The value of the diversity of species is its ability to create a self-reliant and self-contained system that uses very little effort and grows a variety of food.

If you have set up the pond system described in the previous section, with several small parallel ponds for younger

▲ **Chinampa canal.**

fish, or *fry*, you will have a much easier time managing your fish stock. Fish will breed, and your fry will get eaten unless you segregate younger from older fish. Raise the fry in one of the small ponds and release them when they are bigger.

The canal is an even easier system to create, with fry stocked at the source and released into the slow-flowing water later. The chinampa design isn't much different from any other pond construction, with the chief difference being that there isn't much graduated depth. Instead, the land around it is braced with wooden supports so that the edge doesn't erode into the canal. Each "island" is never more than 30 feet (10 meters) wide, and the canal is between 100 and 300 feet (30–100 meters) long on each side. The water level should never reach closer than 3 feet to the garden surface, and the canal should be at least 5 to 8 feet (1.5–2.5 meters) deep. Willows are planted along the banks to prevent erosion. The canal mud is dredged every year and used to grow seedlings, which are then transplanted into the garden bed.

To raise fish, the oxygen level of your pond becomes a crucial factor. Besides keeping the pond clean and clear of too much animal waste and weeds, the aerator can be essential. Solar pond aerators and fountains are available, which can add essential oxygen for longer periods of time than the emergency aerators, mentioned in the previous section, can provide. The biological controls of the pond are in three levels:

1. Aquatic vegetation, like duckweed. These create oxygen during the day and feed your plant-eating fish.
2. The plant-eating fish eat the plants and create fertilizer and debris. Bass or other predator species may be able to live in a netted-off area to control the fry population.
3. Freshwater prawns live on the bottom, cleaning up debris.

You can eat all of these creatures as well. The prawns need help periodically to clean up the muck, which must be dredged up or else it will decrease the oxygen in the water. This precious material should be used as compost or mulch or to grow seedlings as historically was done in a chinampa system.

A pond with lots of fertilizer will grow more algae, and if you want to raise mostly prawns, then you might want to leave more of it on the bottom than usual. Typically, a pH of 7+ is necessary to grow fish, and this is regulated by the addition of fertilizer. A semi-fertilized pond grows tilapia, with ducks and chickens nearby to add a little more manure. A pond with bass or trout must be cleaner. Fish need to eat 1% of

their body weight per day. To grow big and fat, they must eat 3% of their body weight.

Per quarter acre (0.10 hectares) of pond surface, it is possible to raise *all* of this:

40 pounds of bass (predator species)
80 pounds of catfish
120 pounds of bluegill (eats insects)
350 pounds of tilapia (eats plankton and plants)

Tilapia is the most versatile and efficient fish to grow, and for this reason it is incredibly popular. The stocking rates per acre are maximum numbers that are used as a guide by fisheries to show how much you can grow if you were only raising a single species in a pond. See the tables that follow for some general polyculture stocking guidelines. These are some common species used in polyculture.

Tilapia: Tilapia don't like cold temperatures and prefer warm water, but they are fairly hardy and will tolerate most places during the summer. They only take four months to grow from a fingerling to something edible. Duckweed is the best way to feed tilapia. (See the chapter on plant species for more information on growing this amazing water plant). They need adequate shade under water lilies and other plants. To harvest, lower a net into the water and spread worms or breadcrumbs on the surface. When tilapia

▾ Tilapia are delicious.

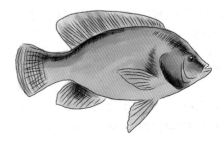

come over to feed, scoop them up with the net. You can winter a few small tilapia in an indoor aquarium or tank. When the weather warms up, release them back into the pond. The stocking rate of tilapia is around 3,000 per acre.

Bluegill: It takes about three years for bluegill to grow to an edible size. They prefer warm temperatures and are often used to feed bass. They eat insects, fish eggs, and small crayfish. They also like vegetation for shade and shelter. The stocking rate of bluegill is around 500 per acre.

Bass: Bass are the most effective method of population control in a pond, because they eat other, smaller fish. They like warm water with lots of plants. The stocking rate is 100 per acre.

Minnows: Minnows eat mostly algae and some insects, and their principle use is as food for bass and other predators. They should be introduced a year before the bass are brought in, so they can build up their size and numbers. They can live to be three years old, but are usually eaten before then. The stocking rate of minnows is up to 2,000 per acre.

Trout: Trout need cold water, and so they do better in big deep ponds. Alternatively, you can raise them over the winter. The water temperature needs to stay below 60°F (15°C) and above 34°F (1°C). They don't do well with other fish besides minnows, and they can eat insects as well, so you wouldn't need as many minnows as you would for bass. The stocking rate of trout is 400 per acre.

Shrimp/prawns: The fewer the shrimp, the larger they grow. Freshwater shrimp need temperatures above 65°F (18°C) but not too hot. They don't survive a cold winter. Shrimp are stocked at a rate of 16,000 to 24,000 shrimp per acre, depending on how big you want the shrimp to get. The fewer the shrimp, the larger they grow.

Mussels: Freshwater mussels prefer a temperature of 50–90°F (10–32°C). They eat very tiny planktonic food at the bottom of the pond and act as a natural filter. You need to use local species that like living in still water. Check your regional fish laws for how many and what type you are allowed to collect, and take a few from several ponds or lakes so that you don't impact the population. They can filter up to a gallon of water an hour, so you don't need that many. They can be stocked at a rate of about 200 per acre as long as they have adequate food to eat.

Catfish: The more catfish eat, the bigger they get, and they eat minnows and small fry. Make sure you have at least 1,000 minnows per 500 catfish. Catfish are winter hardy, although over the winter they don't eat or grow much. Catfish don't do well with other species besides minnows and prefer a clean pond. The stocking rate of catfish is 1,500 per acre.

Crayfish: Crayfish (or crawfish) are another bottom feeder and prefer temperatures around 65–85°F (18–30°C). If the temperature drops below 45°F (7°C) or above 88°F (31°C), they will burrow into the ground and go dormant. They are vulnerable to predators, especially when they molt, because they do come up on the shore. They need lots of vegetation, shallow water, and hiding places. The stocking rate of crayfish is 200 per acre.

Stocking rates depend on two factors: oxygen and food supply. The more you aerate the pond, the more fish you can grow. The more plankton, vegetation, and minnows you have, the more food supply is available. At the same time, the more plants and fish you have growing, the less oxygen will be available and the less you can grow. It is a cycle that you must try to keep in balance. The following tables are illustrations of sample stocking rates in a polyculture system. The rates are intentionally set quite low for better growth rates.

The rules of thumb used here are:
- Reduce everything at least one-third simply to have enough oxygen.
- Avoid growing species that compete for the same food supply, unless you are experienced.

Stocking Rate per Acre

Breed	Monoculture
Crayfish	200
Tilapia	3,000
Bluegill	500
Bass	100
Minnows	2,000
Trout	400
Shrimp	16,000
Catfish	1,500

Suggested polyculture scenarios:

Bass Farming

Breed	Stock rate per acre	Stock rate .25 acre
Crayfish	60	15
Tilapia	1,000	250
Bass	100	25
Minnows	600	150
Shrimp	5,000	1,250
Mussels	60	20

Catfish Farming

Breed	Stock rate per acre	Stock rate per .25 acre
Crayfish	60	15
Catfish	500	125
Minnows	1,000	250
Shrimp	5,000	1,250
Mussels	60	20

IRRIGATION

When you plant lettuce, if it does not grow well, you don't blame the lettuce. You look into the reasons it is not doing well. It may need fertilizer, or more water, or less sun. You never blame the lettuce.

~ Thich Nhat Hanh

What is proper irrigation?

Even a small property can become very self-reliant in the water department with a well-developed water diversion system. It is not only ecological to use your own water sources from your own land—it is more secure. You will have peace of mind knowing where your water comes from and what's in it, and also that you will never be without it due to some disaster. Dams at the top of slopes leading down to dams mid-slope and connected with diversion drains collect rainwater that directs into swales and reserve ponds. The ponds may have irrigation pipes leading to the garden and orchards, which also have swales and diversion drains helping to direct water to the right places. However, an irrigation system should never be used for growing watermelons in the desert or for growing lawns and washing cars. Irrigation is simply a way of supplementing the natural flow of the land and possibly rehabilitating the soil.

▲ An olla, an earthenware pot sunk into the ground used for irrigation.

Desert Irrigation

With some careful planning you can still have enough water to grow food in the desert. All graywater should be recycled to a wetland marsh, and instead of spray or flood irrigation, a drip hose should be used. The drip hose should be buried at least six inches under mulch or soil. Drip hoses can be expensive, but you can make your own version with earthenware pots sunk into the ground (called an *olla*), bottles with holes punctured in the sides, or pipes filled with gravel. In the orchard a sprinkler can be used to spray a mist over roots. Sprinklers are only used in shady tree areas, otherwise evaporation becomes a problem. In dry, hot climates, water only in the early morning, late in the evening, or at night. Mulch and swales are the secret to successful desert gardening, as both of them trap and preserve water runoff.

Designing an Irrigation Systems

Unlike "irrigation systems" that people shell out thousands of dollars for, permaculture irrigation is half species choice and half terrain. Only a small part of it has to do with piping or pumps or equipment. By studying the slope and topography of the land, gravity is used to direct water into the gardens directly or even to individual plants. These individual plants are chosen for their specific traits that make them function well in the location that you want to put them. The barn roof is harvested for rainwater that is stored in a large tank and, hopefully, gravity fed to the gardens below if it does get dry. It is less likely to dry out if the topography and layout of your garden are conducive to water retention, if you have mulched extensively, and if your species can handle a little drought when you can't water them.

WATER PURIFICATION

Drink waters out of thine own cistern, and running waters out of thine own well.

~ Proverbs 5:15

But my water comes from a well . . .

Most ground water is contaminated with a parasite called *Giardia*, not to mention any agricultural chemicals or sewage waste that may have leached into it. Your best bet is rainwater, although it may still have acid in it. Test any water source before you put it in your mouth, and use water purification no matter what. As it will be repeated again in this book, it's also important to store water, even if you aren't using a rainwater system.

The water purification system that you choose will have parts that need to be periodically replaced and cleaned. The water storage tank will also have to be cleaned and the water source tested every year. This kind of maintenance is vital in preventing any serious illness. Testing is as simple as taking a small bottle to your local health unit or lab.

Distillation

The most effective homemade water purification system involves distillation, which is just a fancy name for the process of boiling water to make steam. Dirty water is heated to 212°F (or 100°C, the boiling point) in an enclosed container, it turns into steam which is sent out through a pipe, and the dirt stays in the container. The steam then goes into a cooler where it returns to its liquid state, minus the impurities. The cooler can be anything that can catch the condensation, such as a spiral copper pipe or even a pot lid. Distillation can remove heavy metals, poisons, bacteria, viruses, nitrates, and fluoride. It can't remove oil, petroleum, alcohol, and things that don't mix with water. This is the same process by which alcohol is made, and so you could purchase a regular still. If you are in the desert, you can make a solar still using a sheet of glass to condense the water. Unfortunately, building a still can get you into trouble in some places. It is usually illegal to make alcohol at home, and the act of having a still can become a legal hassle. A solar still does not have the same implications.

Slow Sand Filter

A slow sand filter is simpler but also slightly less effective than distillation. A

▾ Slow sand filter.

▾ Home distillation device.

large-scale version of this is used in most commercial water treatment plants. It is simply a tank 3 to 6 feet deep, with pebbles on the bottom, followed by a layer of gravel, then coarse sand 1 foot deep, and finally fine sand 1.5 feet deep. Water flows through, leaving impurities behind. This method, unlike regular distillation, requires no energy and no pressure. It can remove pathogens and bacteria but has a hard time with heavy metals and dissolved chemicals. To take care of those you can add a carbon filter sandwiched between the coarse and fine sands.

The way the slow sand filter works is not actually through the sand itself. Over time, a layer of scum called *schmutzdecke* forms on the sand. Organic materials get trapped in this green slime and break down. Eventually, the green slime gets so thick that the water has a tough time getting through it, and the filter has to be cleaned. This scum is just scraped out, and the filter is put back into use, but you can't drink the water yet. After two days there should be another layer forming on the sand, and you can start drinking it again. To make this process easier, you can use a fabric filter on the top called *geotextile*, which you simply lift off and clean. This obviates having to replace the sand all the time.

These filters do have a couple of downsides. In cold climates they must be stored inside because they can't be allowed to freeze. If ice forms, you would not be able to clean it, besides the fact that the water would not be able to flow through it. Filters also don't work well if the water is very *turbid,* or cloudy. Dirty water tends to clog up the filter. In warm and humid climates you may have an overabundance of algae. If this is the case, you will have to keep a lid over it. Other than these technical issues, the only routine maintenance involves checking the water flow.

Homes and Shelter

Home ought to be our clearinghouse, the place from which we go forth lessoned and disciplined, and ready for life.

~ *Kathleen Norris*

HOUSE DESIGN

A man builds a fine house; and now he has a master, and a task for life: he is to furnish, watch, show it, and keep it in repair, the rest of his days.

~ Ralph Waldo Emerson

The considerations of a house . . .

Houses aren't just places in which to eat and sleep. In them you can also work, and on a self-sufficient piece of land you may be doing pottery, carpentry, packaging of seeds, computer work, photography, or even health services. Your house will have to implement some space-saving innovations, even if you didn't build it yourself. The bedroom may have to be an office too, herb-drying racks may hang from the ceiling, furniture may fold up against the walls, and storage should exist everywhere. Every inch of space should be utilized.

An energy-efficient home should:

- Never be more than two rooms wide and never more than one and a half times as long as it is wide. In fact, the smaller the house, the easier it is to cool and heat.
- The long side of the house should face the sun, which would be north in the southern hemisphere and south in the northern hemisphere. The bedrooms would be positioned on the shady side, and the kitchen and living room would be on the sunny side.
- The windows and eaves should be angled in such a way that the sun goes directly in and hits heat-absorbing materials such as tile or brick but doesn't come in at all in the summer. This means knowing the exact angle of the sun each season.
- The east side of the house should have some small windows to catch the morning light, but the west and shady sides should have tiny windows, if any.
- Obviously, the house should be extremely insulated. If you are using a concrete slab floor, the outside perimeter of the house should have 2-inch foam stuck 3 feet deep into the ground.

Smart House Features

The entryway is the first concern as you walk into a house. In cold climates, the entryway is where precious heat escapes when people open the door, and it is where dirt and mud appear when you come in after gardening or playing outside. The pathway should be stone or brick, not dirt, to decrease the amount of mud tracked in. The foyer should have a second entry door and have a place

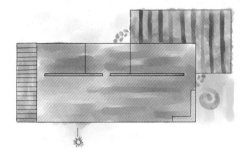

▲ The ideal pantry/mudroom.

▲ The layout of most passive solar homes.

for muddy boots with a drain that can be used to hose down the dirt.

The house is also part of the immediate Zone 1 garden, with vines and plants covering the walls and possibly the roof too. A greenhouse should be built attached to one side, rather than existing as a separate building. A shade patio should extend from the other side. In a hot climate, an inner courtyard with a fountain can also cool a house tremendously. The house isn't just a box for people to sleep in but should instead reflect and utilize the environment in which it is built.

The kitchen will have a work and storage room immediately attached to it. This room may be the most vital area of the house, serving as a pantry for preserved food, with a large tub sink in it for washing produce that is brought in from the garden. There should be a couple of cupboards for drying racks or a place to grow mushrooms. If the kitchen has an exit door, it should also have garden equipment such as boots, tools, and containers for harvesting. If you are designing your own house, it is even better to have the mudroom or foyer lead right into the pantry and lead people into the kitchen.

The Bathroom

In permaculture it is almost always a good idea to have a greenhouse attached

to the house, and the best place for it to meld with your home is in the bathroom. A Japanese style bathroom lends itself particularly well to the attached green house design, it is often already designed with very large windows looking out into the garden and is made of the right materials. The greenhouse can use the moisture and heat generated by the bathing area, and vice versa. The room should be built of stone, brick, or tile to passively absorb the heat of the sun. A Japanese bath saves water but still has all the comfort of bathing.

To take a bath like the Japanese, first you have a quick shower or even a rinse with a bucket. You wash with soap, and have a quick rinse off. Only when you are clean

▼ Japanese bathroom.

The Ultimate Guide to Permaculture

do you get into the large deep tub, which is really more like a hot tub, but not as hot. The whole family often bathes together in Japan, but if you are not up to that, at least the water can be reused because everyone was already clean.

Designing the greenhouse:

When attaching a greenhouse to the side of the house, don't think the greenhouse has to be very big. This greenhouse is not for commercial production. It is usually nothing more than a glass awning coming down off one wall. It is simply an extension of the bathroom or a couple of rooms. The base should be very insulated, optimally built three feet into the ground. There are a few rules of thumb:

1. It should be well sealed, with ventilation at the top and bottom for air circulation.
2. You don't need anything fancy to heat it during the winter. Several 20-gallon water containers stashed on shelves against the wall or under growing benches may be enough mass to keep it warm.
3. Double-paned windows will maintain the heat longer than a single sheet would. At the very least, use wooden window frames rather than metal.

In cold climates the greenhouse becomes the most important part of your home, because it can provide you with fresh food. If the water storage isn't keeping it warm enough, a few bunnies will. You can also use dark curtains or wood shutters at night to minimize heat loss. While it may seem counterintuitive, a vent with an electric fan is also a critical part of any greenhouse. It can very quickly get too humid or too hot on a sunny day, even in winter, and your plants can die.

Designing the shaderoom:

The shaderoom is essentially a fancy patio located on the shady side of the house. It can be designed to have a regular roof, but the walls must be made of sturdy reinforced trellising. These trellises should be grown with vines such as grapes, and a water system can be installed to spray a vine mist. During the summer, if the door between the shaderoom and the house is opened up and the mist turned on, the water will cool the air going into the house. In fact, you can keep a water tank in the shaderoom for passive storage. In a super hot or tropical climate, the shaderoom can extend to porches around the entire house.

Very cold climates:

Heating a house can be costly in places that have severe winters. It is also inconvenient: Snow can damage roofs, condensation on windows can cause mold, and the house can be drafty and damp. In these areas, it makes sense if the house has a small footprint with a steep roof. This means building a house with several stories or attaching it to some other warm building, like a barn. While warm climates have the Japanese bathroom opening directly into the greenhouse, in a very cold climate you should have the option to close the greenhouse off from the rest of the house. In this case a wall partition would be between the greenhouse and the bathroom.

Even then, you can still grow something in the greenhouse in the winter. The fact that the side of the greenhouse is attached to the house for warmth and shelter makes this entirely possible, especially if you keep rabbits or chickens in there for the winter and compost biomass to generate even more heat.

▲ An attached greenhouse can be a beautiful utopia.

▼ Trellising cools a hot area.

The Ultimate Guide to Permaculture

▲ In hot climates the patio becomes a living space.

Very hot climates:

Just as for very cold climates, extra considerations must be made in designing a living area for a hot climate. In many hot climates, palm trees grow naturally and can be used to shade the roof of the house with very little watering. While a small paved area immediately around the house is useful, larger areas should be covered with grass or other green groundcover. Pavement or gravel will simply reflect the heat back onto the house. It is also a good idea to have a separate kitchen, conveniently placed close to the shaderoom. In the summer, a covered barbecue area with an outdoor sink and tables attached to the cool shaderoom can be a pleasant way to spend the evening before going to bed in a house that hasn't been warmed by a hot stove. Hot climates are also sometimes full of bugs at night, but you can leave the windows open at night if you use screens and bed netting. The roof should be painted white to reflect the heat, with a steep angle to allow rain to run off. Some of these hot climates are victim to frequent hurricanes and monsoons, and if so, the house should have deep underground anchors. On the other hand, if this is a desert climate with no rain, a flat roof with a shaderoom and a clothesline might be more appropriate.

Rainwater is even more important in a dry climate. In most deserts the land is formed with steep mesas and canyons because of water erosion, and the valleys are flat and prone to flooding. However, just because the mesa walls are so steep does not mean that you can't still build the house mid-slope. It just means that the house may be set directly into the rock. The water is gravity fed from the dam and storage tank at the top of the mesa down to the cave house.

▲Desert house set into the side of a mesa, with irrigation channels.

Modifying an Existing House

In North America the average affordable home is an older, woefully energy-inefficient, poorly designed leaky box with big windows facing all the wrong directions. Even this typical house can be improved without any major construction:

1. The easiest way to improve a house is to caulk and seal all the windows and doors. If you can afford it, install new double-paned windows, but even sealing all the cracks will make a huge difference.

2. While it may be more difficult to add insulation to the walls, it may not be so tricky to get into the attic and add insulation to the roof, which can make a huge difference. You could rip out the drywall and add more insulation to the inside (which would be a major job), but it is easier to wrap the outside of the house.

3. A greenhouse, even a tiny one, can be added to almost any house. Ideally, the greenhouse should be added along one whole wall, and if possible at least part of it should open into a stone or brick bathroom. If ripping out a wall is not possible for you, you could add one

with at least a door into the house. In a cold climate, a doorway that can be closed off is a better option anyway. At the very least, a greenhouse attached to a couple of windows would be an improvement. If you can't even do that, a skylight could be an option.

4. The greenhouse should be built with concrete, brick, or stone to act as passive heat storage. If your house happens to face the right direction (north in the Southern Hemisphere and south in the Northern Hemisphere), you might be able to build a brick or rock wall inside facing the windows for another passive heat storage. Most houses have a large living room window that allows sunlight in, with a wall separating it from the kitchen. That wall and the living room floor can be covered in tile.

5. Attaching a shaderoom should be something that is possible for most people. It should be built on the shady side of the house and have a doorway into the house from it so it can be opened up for a cool draft. Apartment dwellers already have this in the form of a patio, and most people are able to grow plants on it, which can serve the same purpose.

6. Unnecessary appliances should be eliminated, and essentials should be switched to energy-efficient and passive solar options wherever possible. A solar hot water heater can supplement the hot water tank, and you can also switch that out for an on-demand hot water heater. Drying racks and clotheslines supplement or replace the clothes dryer. When shopping for energy-efficient appliances, keep in mind that North

The Ultimate Guide to Permaculture

American brands are pathetically behind European standards. Many *EnergyStar* appliances don't even come close to most European-made appliances. These brands might be more expensive, but the energy and water savings are so extreme that purchasing these brandsis worth it.

The Cellar or Cold Storage

A cellar is an underground room that stays cool and dark and isused to store food. In the old days, most houses came with a cellar for storing food and staying safe from tornadoes. The cellar had an outside door and was unlikely to be attached to the house. Today a cellar is still used as a storm shelter, but it is often located in an unheated area of the basement. It should be designed to maintain a temperature 32–40° F (0–4° C), the average optimum climate for most food storage. If you decide to dig a cellar detached from the house, you have to be careful not to hit any pipes or wires.

- Make sure that your water table will be lower than the floor throughout the year or you will have flooding. You may need to build extra drainage around the basement or cellar just in case.
- Don't use pressure-treated wood. Pressure-treated wood resists rot from moisture and humidity, but you can't use it on interior framing or in contact with masonry.
- Use stone or reinforced concrete to support an underground room.
- Leave the floor dirt in the cellar. Cement will crack and absorb moisture, and for food storage moisture in the air is a good thing. This kind of floor can make it difficult to resell your house if it is part of the basement, so if you do want to have a dirt floor, the cellar might have to be an extension of the basement or a separate entity.

▲ Clotheslines are a no-energy system.

BUILDING MATERIALS

Humans need continuous and spontaneous affiliations with the biological world, and meaningful access to natural settings is as vital to the urban dweller as to any other.

~ *Dr. Stephen Kellert*

Structural Materials

Bamboo: This swift-growing tree is actually a type of grass. The wood can be used to build just about anything. It is traditionally used in very warm or tropical climates, but it is not very rot-resistant, so it must be set on an adequate foundation that keeps it away from the ground. In North America it is more common to see bamboo used as framing combined with adobe or cob walls.

Wood: Wood is the main structural material in most North American homes. If you decide to use your own trees, use old growth, not young trees. The young trees are valuable because they have already become established and can repopulate the forest.

Adobe or cob: Adobe is a mixture of dirt and other materials into a kind of clay that is packed very hard. It has traditionally been used in hot desert climates, and the building process involves making hard clay bricks, which are stacked like any brick house and plastered over with another layer of adobe mud. Cob is very similar but has a different mixture and has been used all over the world. The building process of cob involves building a framework of wood, and filling it in with sculpted piles of mud clay.

Dirt: Dirt homes are either dug into the ground or formed by stacking dirt to

▲ **Bamboo home.**

form walls. In the old days, strips of sod were simply plowed up and used like bricks, or a little dugout was hollowed out of a hillside, but today dirt homes are very sophisticated. Underground and earth berm houses are very popular, but an even easier option is the earthbag. Earthbags are filled with dirt and stacked with a rebar framework.

Canvas: Yurts and tipis are cool in the summer, and if they are built right and used correctly, they can be very warm in the winter. Yurts are more often used as year-round houses with commercially made kits available that provide all kinds of options and insulating capacities.

Stone: Most places have lots of stone available, which can be used to build a well-insulated house. Stone houses may not work so well in an earthquake zone, but they are the best option in a place that sees frequent destructive storms. It is time-consuming to build with stone, but the resulting structure is extremely long-lasting.

Strawbale: A modern popular alternative building method is to simply use bales of hay. These bales must be tightly twined and are held together with a rebar framework. While straw itself isn't the best insulator, the bales are thick enough to provide excellent insulation.

Cordwood: Cordwood is an old building method that has been made popular again by natural homebuilders. Rather than using whole logs to build a wall, logs are sliced into circles or disks and stacked with cement to hold them together. This method is useful where wood is scarce, and it still offers better insulation than many modern building methods.

Types of Insulation

Dirt: Dirt is a terrible insulator. At only R-0.25 per inch (2.5 cm), you might wonder why so many homes are made of dirt. While it is terrible at insulating, it is also terrible at transferring heat and cold. Because of this,

it is a fantastic thermal building material, and self-regulates the environment over 12-hour cycles, which is precisely the cycle of the sun. This means that during the day it will radiate the coolness of the night and at night will radiate the heat of the day.

Air: Air is a decent insulator, if it is sealed off. It has an R-value of R-1 per inch. Usually it is used to insulate windows, which is why double-paned windows are better.

Sawdust: Sawdust was used in the old days in icehouses to preserve ice over the summer because it is an effective insulator. It should be sealed in bags of plastic, and it needs a vapor barrier as well. It only has an R-value of about R-2 per inch, so the thicker the better.

Balsa: Balsa grows quickly, and the cotton from the seedpods as well as the wood can be used as insulation. Pressure treated balsa wood varies in R-value, but the less chemicals it retains the better job it does. Its R-value is between R-2 and R-3 per inch.

Paper: Soak shredded paper in one part borax and ten parts water to reduce the flammability. Commercially available cellulose is essentially the same thing, with a few more chemicals in it. The R-value will be between R-2 and R-3 per inch.

Straw: Straw is another material that has been used for centuries. While structural insulated panels are popular and often made of processed wheat straw,

it is more common for do-it-yourself to use straw bales. Straw has an R-value of R-2.4 per inch.

Cork: This amazing sustainable wood product is available in a variety of commercial forms, including slabs, tiles, and blocks. It comes from the bark of a tree rather than the core, so the tree stays alive and regrows fairly quickly. It has an R-value of R-3 per inch.

Cotton: Commercial cotton insulation is actually a mixture of recycled cotton and plastic fibers that have been treated with flame retardant. It is sold in rolled batts that have an R-value of R-3.4 per inch.

Hemp: This plant should be better known and used more. It is a versatile, quickly growing plant that resists decay and rot. It is commercially available in rolled and has with an R-value of R-3.5 per inch.

Plant fiber waste: Any plant fibers can be used as insulation, and if matted together and made into a batt, they would have a similar R-value to hemp and wool insulation.

BUILDING WITH DIRT

I bequeath myself to the dirt to grow from the grass I love,
If you want me again, look for me under your boot-soles.

~ Walt Whitman

Underground Houses

Underground houses have been built of stone and dirt in every climate for thousands of years, completely submerged in the ground or in hillsides. While dirt isn't necessarily a great insulator, the sheer quantity of dirt used for an underground house acts as a thermal mass, keeping the house warm in the winter and cool in the summer. Most modern North American underground homes are actually *earth berm* houses or are covered in a man-made embankment. The front of the house is shaded with vines, and the roof can either be part of the berm, with plants growing on it or not. In the desert these houses are set far into the soil or even into rock faces, while in more temperate climates only the back half of the house may be earth sheltered.

The simplest and most fool-proof method of building an earth berm home is to build the house with steel framing or reinforced concrete, and in fact in many places this is the only legal way to do it. The house is built with either a circular or dome shape for strength, It's not clear what is built small. in the parts of the house that are covered in dirt. The enormous weight and pressure exerted by all of that soil could quickly crush you if the house is designed poorly. Inexpensive pre-designed plans are available, and they are worth the investment.

Besides the structural strength and engineering considerations, the home also has to be very waterproof because of all the moisture in the soil. This includes polyethylene sheeting, such as pond lining, clay, or various commercial cement products. A layer of insulation is added on top of this waterproof membrane, outside the house, to protect the plastic or concrete from freezing or damage. This is usually made of polystyrene sheets or spray foam. Some people don't insulate the inside walls of their underground home, but it is a good idea in places where the frost line extends deep into the earth.

While earth is a more environmentally friendly material, you should note that the amount of unsustainable materials that is needed to make an underground house safe and livable is quite large. There are other options for amateur builders that are easier, cheaper, and more sustainable.

Earthbags

An alternative to earth berm houses is the popular and extremely low-cost earthbag method. Rather than building traditional framing to hold up the dirt, the house is built with stacked bags full of soil. This alternative creates thermal mass and is incredibly strong and long lasting. You can simply cover the earth bags with adobe or plaster. The bags can be recycled sacks, misprinted brand new bags, or burlap, but polypropylene is stronger and lasts longer. They should be able to hold 50 pounds, at least 17 inches (43 cm) wide and 30 inches (76 cm) long. If you are using bags of that size, you would need approximately 140 bags per 100 square feet (9 square meters) of wall. A

The Ultimate Guide to Permaculture

Underground home with large windows.

thousand bags might be able to build 700 square feet (65 square meters). The shape of the house can be almost anything you can imagine, and many builders create spheres or even cones.

It's important to note here that while mixing an earthbag with an earthberm house may seem like a good idea, there hasn't been much experimentation with earthbags under earth berms. It requires extensive knowledge of engineering to predict the pressure of the earth on the bags. A circular house is stronger, but since the walls are simply stacked, the house could not be very big if you wanted to pile dirt on it. The R-value of earthbags alone is already enough to create a super-insulated house without putting it under a hillside.

The basic building process steps are:

1. An earthbag house does not need a traditional foundation, although you may decide that you like a poured concrete floor. You first need to excavate a level area, which you can use to fill your dirt bags. You may also decide you want a basement, in which case you will need to use reinforced concrete and excavate a large area. That gives you even more dirt for your bags.

2. Dig a trench the width of the base of your wall, which follows the outline of your house. This is going to be your footer. The width of the base of the wall must be at least a tenth of the height of your wall: I your wall were 10 feet high (3 meters), the base of your wall would have to be 1 foot (30 cm) thick. Trenches in wet and cold climates need to be dug below frost level.

3. Fill the trench with gravel or concrete. If the ground isn't super sandy, you can use gravel. If the trench has been dug below frost level, you don't need to fill it up to the top. The first layer of bags can be filled with gravel and laid in the trench. If you have a very wet climate, the first few rows of bags can also be filled with gravel. Some people have also embedded rebar in the concrete to pin the base down.

4. If you live in a very wet climate, you can also install a perforated pipe in a trench dug around the perimeter of the walls. The pipe is set down on a bed of gravel. You can fill that trench with regular soil. You could also build a swale beyond that to force water into the soil to be soaked up by trees rather than by the foundation of your house.

5. The bags must be filled with a mixture of materials, hopefully from the land on which the house is built. The mixture should never have more than 30% clay and shouldn't have too much organic material in it, but generally, whatever dirt you have should work fine. You can even use waste rock or gravel, volcanic rock, crush shells, or any mineral that compacts very tightly when compressed. The best material is a mixture of sand and clay. Some people try to use topsoil, but since you need this valuable humus for growing food, it doesn't make sense to use it for this.

6. When you fill the bags, you don't need to seal them unless their ends are

exposed. Normally, you can just fold the top over when you lay the bag it down. If you do need to seal it, use a stapler.

7. The bags must be tamped down to compact the materials. This is the most labor-intensive part of the process and also requires the least thinking. Usually, a long-handled tool with a flat plate on the bottom is used to smash the bag as flat as possible. This is a very important step; without it, the bags might shift.

8. On top of each row of bags, you must lay two strands of four-point barbed wire. The wire will embed into the bags and hold them steady under tension or in an earthquake.

9. Windows in earthbag homes are usually triangular because that is the easiest shape to make without adding support. If you do want square windows, the *lintel* or support above the window must be able to hold the massive weight of the earth above it. A metal or concrete base is more likely to handle the load, or otherwise your wood beam would have to be massive.

10. Interior walls and hallways are often built with archways in earthbag homes, to avoid using wood. Arches are constructed by *fanning* the bags, or sandwiching them together until the peak bag is propped vertically. This is done using a form made of wood that holds the structure it together while you are building it. Since a doorway or hallway is near the top of the house, it doesn't have that much weight on it, and using a nice wood beam is possible.

11. Piping and electrical wiring can be installed inside wood framing in the various rooms that need it. The design of the house should be conducive to this by placing the bathroom, laundry, and kitchen back to back, and having

▲ **Archway built by fanning bags.**

all of the electrical components placed centrally so that wiring doesn't have to go very far. The piping from the water source outside can come in through the floor or the roof. Woodstove piping will still burn bag material and still needs to be properly insulated, as it would be in a regular house.

12. The walls must be built and covered fairly quickly. More than a month of intense desert sun will degrade the bags too much and render them unusable, and no more than three months in a temperate climate will produce the same effect. Once the walls are built, they need to be covered in a moisture barrier and then smeared with a plastering material. People have used cob, cement stucco, and lime plaster to cover their walls. Cob is a mixture of clay, sand, water, and straw in varying ratios to produce a sticky, globby material that is spread by hand and hardens into an impermeable brown layer. The mixture is about one part sand, one part clay, and 10% straw combined with just enough water to hold it together. To make lime plaster, mix Portland cement, lime, and sand in a ratio of 2:1:9. The most durable and waterproof solution is cement stucco, but it is more expensive. To apply any plastering material to the wall, first wrap the whole house in chicken wire and smear on a 1-inch (2.5 cm) thick icing.

▲ Earthbag houses can be covered in cob.

13. The roof is attached with embedded rebar rods that hold a wood header. It is not a good idea to build more than one story because you lose stability in the walls. A dome or cone roof looks unique and saves having to buy roofing materials but is very prone to deterioration due to moisture. Even cement stucco is likely to leak if it is directly exposed to the elements. It is also very difficult to harvest rainwater from such a roof, and for this reason a traditional peaked roof is recommended. The eaves should extend far enough so that the walls of the house wouldn't be exposed to moisture at all.

OTHER DIRT BUILDING METHODS

Every book is a quotation; and every house is a quotation out of all forests and mines and stone quarries; and every man is a quotation from all his ancestors.

~ Ralph Emerson

Concrete

While concrete isn't the most ecological option, it is one of the simplest tools for the amateur homebuilder and is also relatively inexpensive. In the grand scheme of things, using concrete in conjunction with green building materials like rocks or cordwood is still much "greener" than traditional building methods and more energy-efficient.

A stone house is a time-consuming but worthwhile endeavor, especially if you have a property with lots of rocks. As you go through your garden, weeding or planting, you will inevitably come across many stones, which you can set aside in a rock pile for a house project. There are two main building methods: the old-style stack and mortar method, and the slip form method. First, you must dig a trench for the footer and fill it with concrete, or pour a concrete slab. The stack and mortar method simply involves laying the rocks along a straight line, carefully stacking them together, and sometimes shaping them with a hammer. The slip form method involves building a wood framework that holds the rocks in place while the concrete hardens. The flat side of the rocks is pushed up against the outside wall for aesthetics, and concrete is poured in behind. When the slip form is removed, the wall has beautiful flat stonework. The wall should be 8–12 inches (20–30 cm) thick, with a ½ inch (1 cm) of mortar between each rock.

Cordwood is another energy efficient and relatively easy construction method. Round discs of wood, called *face cords*, at

▾ Cordwood house.

▲ **Slip form concrete house.**

least 6 inches (15cm) wide are stacked and sealed together with concrete. The wood must be debarked and dried so that there won't be any shrinkage. If the wood shrinks after it is set in the wall of the house, it can create holes and let bugs in. At the same time if you let it dry too long, the wood can split, which you don't want either. The wood is stacked to form a wall the same way that a stone house is built. The face cords are set on their sides so that the outside of the house looks polka-dotted. The wood lasts longer if it is not sealed.

Slip form and cordwood are both simple, cheap, foolproof construction methods that require little engineering knowledge and yet make beautiful, durable homes. However, they are much more time-consuming than other methods. If you must choose between the two and have a supply of rock, choose slip form. It looks impressive and lasts even longer than wood. You don't have to do much finishing

work on the inside; covering the beautiful rockwork with drywall would be a shame. Some insulation can be added to the outside of the house, covered in stucco or wood siding.

Sod roof and plant integration . . .

A sod roof gives you the option of having a living roof without building an underground house. To be covered in dirt, your roof must be very strong and well supported. A moisture barrier must be placed between the roof and the sod, and a beam along the edge of the roof keeps the sod from sliding. A sheet of metal hangs off the eaves and rolls under to save work cleaning the gutters. In a wet area, regular grass works well, but in other climates (such as the desert), a local groundcover species might work better. In some places bulbs and herbs can also work. A watering system fueled by rainwater keeps the plants growing during a dry spell in the

▲ Living roof.

summer. If a sod roof is too heavy for your roof, try growing ivy. The ivy will spread and grow over your entire house.

SUSTAINABLE LIVING HABITS

What's the use of a house if you haven't got a tolerable planet to put it on?

~ Henry David Thoreau

House temperature strategies:

1. All windows should be covered in heavy curtains from floor to ceiling. In the winter these are shut at night, and in the summer these are shut during the day.
2. In the summer, keep the windows open at night and shut them early in the morning before the sun starts to warm up. Bamboo blinds on the outside of the windows can make a huge difference during the day.
3. The shade room can be opened up into the house to create a cooling draft in mid-summer. Open all the doors in the house, and open the vent at the top of the greenhouse to let heat escape. The draft will sweep through the house, pushing the hot air out.
4. A ceiling fan can be enough to create a draft on a still, humid day, without using as much electricity as an air conditioning unit.
5. Weather stripping should be installed around doors and windows before winter.
6. The tile floor and brick wall act as heat banks in the winter if the sun is allowed to hit them during the day. At night they will radiate that heat back into the house. In the summer the opposite can be achieved by letting cool air hit them at night, and they will radiate coolness during the day.
7. If you want to cook or bake, do it at night and prepare things the day

before you need them, or cook outside. A hot stove can heat up the house tremendously.

Recycling

The first step to recycling is avoiding. A typical North American person makes 5 pounds (2.2 kg) of garbage *per day*. Most of this trash goes to landfills, and a small part of it is recycled. This is mostly made up of plastic bags and packaging, food waste, paper from packaging and labels and office waste, and cardboard. Some of it is aluminum cans, glass jars, and little items like bread tags. With planning, it is possible to make little or no garbage at all.

Plastic bags: There are several kinds of plastic bags: grocery shopping bags, thin produce bags that use twisty ties, slightly thicker product packaging bags (often stapled together), etc. Use cloth shopping bags and mesh produce bags. Keep other bags to store leftovers or other items or cut them into thin strips for crochet projects.

Food waste: This should be composted, either in a compost bin or in a worm bin. See the following section.

Paper: Office waste and scrap paper can be ripped up or shredded and added to any compost or worm bin. There has been concern about shiny paper: receipts, magazines, newspapers, can labels, and other slick paper items, because the inks contain such toxic substances, but today they are much less harmful and compostable in small amounts. Receipts are now the only paper you can't do anything with, because they are coated with BPA. Magazines should be shared through a free magazine rack or given to friends first, or cut up for collages and art projects. The little bits that are left after all of that can be composted.

Containers: Glass jars and bottles should be washed carefully and stored with their lids until you are ready to use them again. Beads, buttons, bread tags, bottle caps, and beans all store well in glass. Plastic tubs and containers with lids can be used to store leftovers, while aluminum cans are used to store fat drippings for making lard, punctured to make lanterns and garden-waterers, or put aside for string telephones.

COMPOST

Truth is the strong compost in which beauty may sometimes germinate.

~ Christopher Morley

Kitchen Compost

Food scraps and kitchen waste should be added to the compost heap, where they will decompose and can eventually be added back into the soil from whence they came. For most people that is about 30% of their garbage. Keep two buckets under the kitchen sink and add all of your food scraps *except* meat and dairy. Orange peels, banana peels, and eggshells should be crushed and cut up so that they will break down faster. You can also add small paper scraps. Once a bucket is full, dump it out in the compost heap. There are commercial products made of black plastic that have convenient lids and look nice. These products work well in an urban environment because of their size and aesthetics; they absorb heat because of the black coloring, helping your compost break down a bit faster. However, if you are using a composting toilet, you'll need a much larger bin. If you are placing the compost in the greenhouse as thermal

▲ Highly efficient compost bins.

mass, it should be at least 5 feet (1.5 meters) in diameter. If you are building wooden bins outside, they should be at least 5 feet (1.5 meters) squared. Every material that is not going to be used as mulch should be put in the pile, including food scraps, composting toilet solids, leaves, garden waste, and manure, and it is a good idea to collect material from elsewhere, such as seaweed or organic material from commercial garden services (although you would need to make sure it was not treated with chemicals). Mulching items like leaves, straw, pinecones, bark, sawdust, and other materials in large quantities are kept separate. The pile must be sheltered from the rain unless you live in a very dry climate.

Once you've collected all of this material together, it will begin to heat up, usually within 24 hours. Organisms will gather to feast on the materials you brought for them, which will make the temperature rise. During this time it is best to leave the pile alone and just keep adding more on the top. It should rise over 120°F (49°C). When these microorganisms are done, the temperature will drop, and you can stir the pile to get some oxygen mixed in. Insects will move in, or you can start raising worms in it.

Sometimes compost piles don't heat up, which is fine. It just takes longer. This is usually because the pile hasn't gotten bigger than 3 feet square, or it might be too wet or too dry. It should be about as moist as a damp sponge. Another problem that often arises is an imbalance of materials. Woody materials (called *brown* materials) should make up about half the bulk. This includes paper, leaves, dry grass, and straw, which are high in carbon. The other half is *green* materials, or things that are high in nitrogen, like manure and garden waste. This is not an exact science, and you don't have to measure it, but experience

Compost, mulch, or animal food?

Compost	Mulch	Animals
Banana peels	Cardboard	Dairy products
Ashes	Pine needles	Diseased plants
Hair	Fabric	Fish
Citrus peels	Grass clippings	Meat
Coffee grounds	Hay	Bones
Corn cobs	Straw	
Dust	Leaves	
Lint	Nutshells	
Eggshells	Newspaper	
Manure	Sawdust	
Soil	Weeds	
Toadstools	Wood chips	
Rhubarb stems		
Seafood shells		
Sod		
Veggie soup		
Tea leaves		

Note:

Corncobs and peels should be chopped first.

Rhubarb leaves are poisonous, so make sure they really get cooked.

Crush seafood shells first.

Sod should be placed upside down.

Use untreated sawdust only.

will begin to tell you what you need more of. Often if the pile is too wet, it needs more brown materials, and if it is too dry, it needs more green materials. It helps to chop up all the materials before adding them, so they'll break down faster.

Worms, or Vermicomposting

Raising worms is no more difficult than composting. It is especially valuable to urban dwellers and can even be done in an apartment. The worms need a covered bin with holes in the sides and bottom. If you want, you can line it with a mesh fabric to keep the little ones from crawling out. Set the bin on a drainage tray because you'll be watering it like a houseplant.

1. Fill the bin with a bedding material. You can use nice black soil, shredded newspapers, or a mixture of both.
2. Now go out and get your worms. You need the kind that is sold for fishing and not the kind that is probably in your garden already. These are called red wigglers, redworms, or branding worms. You will need about 0.5 pound (0.23 kg) of worms, or 500 regular sized worms per cubic foot (0.03 cubic meters).
3. The worms will eat their weight in kitchen scraps *per day*. Bury your leftovers in the bedding after each meal and follow the same rules as for your compost pile: no meat or dairy products. Garlic and potato peelings are also a bad idea. They do need a steady diet of crush eggshells sprinkled on top.
4. They don't like cold temperatures. You can keep them outside in the summer, but if the nights start getting chilly, bring them in.

▲ **Worm bin made out of plastic tubs.**

5. Whenever you take the lid off the bin, the worms will dive down to the bottom because they hate the light. Every couple of months open it up and put them in some bright light. After ten minutes you can scrape off the top layer of materials, which is all of the valuable worm castings that are great for the soil. When you see worms, wait another 10 minutes and do it again. Repeat until you've gotten it all, and fill the bin up with fresh bedding again.

Comparing Composting Toilets

A typical suburban or rural home is hooked to city sewage or to a septic system. A septic tank is a concrete or plastic container set in the ground, which holds your sewage while it allows it to slowly seep out into the ground through a leach field. The average toilet uses 3.5 gallons (13.25 liters) of fresh water per flush to facilitate this process. If you use the toilet three times per day over an entire year, this fresh water can really add up: 3,832 gallons (14,505 liters). It is for this reason that a composting toilet makes so much sense. The actual compost harvested from a family will not supply an entire garden, but the water savings is tremendous.

It is also now possible to buy beautifully designed composting toilets at a much lower cost than a septic system. There are many composting toilets on the market, but not all of them *compost* as claimed. For an average family, the cost of a toilet that does all the composting in the same container would cost many more thousands of dollars than is necessary. These involve a large containment bin located in the basement of the home and usually have a heating element to evaporate the liquids. Other, more affordable toilets claim to facilitate the composting process in the containment bin but actually can only be used on weekends in vacation cabins. Both of these are prone to several problems besides their prohibitive cost. They tend to overestimate their capacity, which can leave the owner with a soupy (and disgusting) overflowing mess, they can sometimes breed flies in the house, and their ugly bulkiness makes them difficult to install in most homes.

The best design is a waterless system, which separates the liquids from the solids and utilizes a powerful fan to remove odors. The solids go into a compostable bag in a large bucket, and the liquids are diverted to graywater separately. The graywater system can be designed to deposit directly into a garden. The solids must be carried out of the house every few weeks for decomposition outside, but once the waste is safely composted (a process that takes about a year), it can be added to the gardens. These toilets are often much less expensive than the self-contained versions and are also much smaller, so they can be installed anywhere.

Humanure Toilet

You can build a system similar to the waterless toilet by using a simple bucket, peat moss, and sawdust to keep odors away. In the old days, and even in some rural places today, the traditional low-tech solution was an outhouse, which is simply a pit with a tiny shed built over the top of it. In the shed is a hole with a seat. These pit toilets are highly regulated, and if you live anywhere near civilization, they are probably illegal. *Humanure*, on the other hand, is odorless, doesn't leech into the groundwater if done properly, and is much easier to set up. To build your own humanure toilet:

1. Get a 5 gallon (19 liter) bucket and fit it with a comfortable toilet seat that can be removed. The simplest way to do this is to build a wooden box with a hinged lid, cut a hole in the lid the size of the seat, and screw the seat to the top. The bucket just sits inside the box for easy emptying.

2. Keep a supply of sawdust next to the toilet, and every time the toilet is used, add a thin layer on top of the waste you just added. This is enough to keep any smell down. The sawdust should come from a sawmill that is only cutting logs, not pressure-treated lumber. The pressure-treated wood is filled with toxic chemicals that you don't want to eventually be adding to your garden.

▾ **Separating composting toilet.**

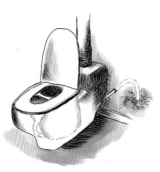

A note about smells and other nasties

Will it breed flies? Will it smell? Will it harbor diseases? Will it bother the neighbors?

A properly managed composting toilet will do none of these things. It does take more effort than a flush toilet, but the principles are simple:

- Empty the humanure from the house frequently
- Make someone in charge of ensuring it stays covered with dry material
- Make sure your compost pile gets *hot*
- Bring in dry materials from somewhere else for your pile if you need to
- Build your pile on a concrete slab
- Put it inside the greenhouse or build nice wooden walls for it

3. Once the bucket fills, it must be dumped in an outside pile. The compost pile needs to have a concrete base covered in a large layer of soil with a layer of leaves and grass and other absorbent material on top. The concrete is essential in order to avoid leeching. This pile can be put into the greenhouse for thermal mass, as described in other sections. Or you can simply put it in a plastic bin on top of a small concrete slab just like regular compost, but if you choose this route, the bin should not be airtight because oxygen is necessary for decomposition.

4. A composting toilet does need a little more effort than a regular flush toilet. The lid should stay closed to keep flies

▲ Sawdust toilet.

away, and you have to remember to keep that sawdust supply well stocked. The bucket must be emptied before it gets too full or waterlogged, and it should be taken out immediately if it starts breeding flies. An average family shouldn't have to worry about flies too much because you will be dumping the bucket frequently enough to prevent problems. Every so often the compost pile must be raked flat, and you can add other waste to speed the breakdown of materials.

5. When the bucket is full, dump it in the compost area and cover it with a layer of woody material. There is a strategy to this. Keep a small shovel nearby so that you can dig a small hole in the middle of the pile. Dump the bucket and then cover it back over with the stuff you dug out. Cover that with more dry material like straw or leaves. Every so often the compost pile must be raked flat, and you can add other waste to speed the breakdown of materials, but don't turn the pile. Turning it will only cool it down. Clean out the bucket with a little water and some biodegradable soap and *dump that water back onto the compost pile* so that you don't contaminate anything. Add a layer of sawdust at the bottom

of the bucket and put it back in the bathroom.

6. The compost pile does need to get hot in order to kill bacteria. The temperature should rise above the temperature of the human body (98.6°F or 37°C) for an extended period of time, and hopefully above 49°F (9°C) for at least one day. If it is too cold, add more organic materials to help get the decomposition process going. There is no rush to this because the pile has to be quite big for it to happen, but it is the end goal. You will notice that this composting system is different from the waterless commercial toilets in that the liquids mix with the solids. Since you must add so much dry material, the liquids are necessary to help keep the compost at the right balance—not too dry, not too wet. If your pile never gets hot enough despite all you do, it has to sit undisturbed for two years. After that you can use it.

GRAYWATER

If there is magic on this planet, it is contained in water.

~ Loren Eiseley

What is Graywater?

Graywater is the water that drains from sinks, tubs, washing machines, and showers. If you use a compost toilet like the waterless one described in the previous section, then you might also divert the liquid wastes from that as well. Graywater is different from *blackwater* in that it doesn't contain sewage: solid human waste. On most systems the graywater is simply mixed with blackwater and dumped into a septic tank or sent to the city sewer. This is obviously a waste. Graywater, if handled properly, reclaims valuable nutrients and puts them back into the soil, while at the same time saving energy and water.

Keep in mind that graywater systems might be illegal where you live, although is changing. Where building codes do exist to regulate them, you will have to get the system inspected and approved.

Graywater cannot be spouted directly into a food garden, because it will contaminate your food. It can, however, be piped into a marsh or container garden where it will help grow a variety of water-loving plants. The first consideration is location. Either you need to use an existing marsh, or you will need to pipe it somewhere where nothing important will be contaminated. This means a place that won't flood, isn't near any food, and where the graywater won't harm any important species.

1. From the graywater source, install a 2-inch (5-cm) pipe from the house to the graywater treatment area with a grade of 2%, put through a trench on a bed of sand. Install a shutoff valve and two screen filers: one on the inlet and one on the outlet. Fill the trench with soil.

2. Calculate your water usage. A frugal household of five with a waterless composting toilet and no dishwasher uses about 940 gallons (3,558 liters) of water per week.

3. Your marsh area should be able to handle the volume of water that is entering it in a day. One cubic foot (0.03 cubic meters) of wetlands will filter roughly 1 gallon (4 liters) of water. Your marsh surface area should be able to handle about a third of your daily graywater production, and the marsh should be 2 feet (60 cm) deep.

Here's the formula:

(**Daily graywater** in gallons or liters)/ 3 = surface area of marsh

Example:

134 gallons per day/3 = 45 square feet

This could mean a long skinny marsh 2 feet wide and 22 feet long (0.6 m x 6.7 m) or a small area 5 feet wide and 9 feet long (1.5 m x 2.7 m). In a rural environment the marsh would be placed in Zone 2 or at the edge of Zone 3. In an urban environment an alternative method is used. Rather than building an open marsh, the pipes run into a container garden. This container garden would have the same capacity and organic materials as the marsh but would have to be formed of either one or more concrete planters placed in Zone 1.

4. Excavate the marsh or build your concrete containers. When you dig, angle the slope away from the graywater pipe so that the water will flow over the entire marsh area. Pile the dirt around the sides as a retainment wall, and plant the wall with clover. The inside of the marsh or planting container should be layered with a variety of filtering materials, starting with a layer of gravel, then sand, then

▾ Marsh excavation.

a layer of sticks cut into 6-inch (15-cm) lengths, and finally a layer of straw and other mulches. The mulch must be at least 8 inches (20 cm) deep and will have to be replenished every year.

5. Now you can begin growing wetland plants in the mulch. Local varieties of cattails, rushes, reed grasses, horsetails, and water-loving flowers are good choices. The cattails are edible, and rushes and reeds can be used for making baskets. These plants are also part of the filtering system, purifying the water as they take it up into their roots.

6. There are a few rules to using graywater. You must use biodegradable soaps and avoid any detergent. You also cannot wash anything with human waste on it, such as cloth baby diapers, and send the water to the graywater system. If you have diapers to wash, you will have to have a way of keeping this blackwater separate. You could have a valve for the washing machine that sends water to the city or septic if you need it, and then switches back for other loads. Or you could give them a rinse and wash by hand and dump the water with your humanure compost before putting them in the washing machine.

7. The water level in your marsh should never flood over the top of the mulch. If you find that you are over capacity, you can install a tank with a valve to control the flow of water, or increase the size of your system. If you do use a tank, the liquid in it must cycle every 24 hours or it will be too toxic to use anymore. Watch for clogs and prevent them by flushing the system with clean water once a month.

LOW ENERGY REFRIGERATION

The same system that produced a bewildering succession of new-model, style-obsolescent autos and refrigerators can also produce an endless outpouring of new-model, style-obsolescent science.

~ Harvey Wheeler

Freezer Box Fridge

A practical alternative to a regular fridge is difficult to find. Even the most energy-efficient models are woeful energy hogs, and yet the capacity is so necessary, especially when you are growing so much fresh food.

The simplest and easiest solution is to take a chest freezer and replace the thermostat with a regular refrigerator thermostat. Cold air is heavier and tends to fall, which means that every time you open your typical vertical fridge, all of the cold air falls right out the door. A chest freezer has the door on the top to prevent this inefficient escape of cold air.

Freezers

Freezing food is a quick and healthy method of preserving food, and even though a freezer requires electricity, it is still worthwhile to have. The best temperature is -5°F (-20°C), but to save energy you can go as high as 0°F (-17°C), although that's pushing the limit for keeping meat safe. Keep a few plastic jugs of water in the bottom of your freezer, so if the electricity goes out, the food will last longer and you will have a small water supply. Keep the freezer in the coolest room of the house, but not in a place where it could freeze. Freezing temperatures kill the motor.

Non-Electric Refrigeration

If the electricity does go out, keep the door of your fridge closed and cover it with blankets, except for the motor vent, if you know the power is coming back on soon. If you know the power will be out for a while, you will have to pull everything out and build an outside refrigerator using a large camping cooler (an "ice chest"). Dig a big hole in the ground, stick the cooler in, and stuff insulating materials like straw and bricks around it. Cover it up with something heavy so animals can't get in. Keep in mind that meat won't last long in the cooler. Use a fire or barbecue to cook some of the meat that you plan to eat the next week. Cooked meat will stay good in fridge much longer than raw meat, about five days rather than three. The rest of the meat needs to be salted and dried.

A more long-term and simple solution involves two large clay pots or brick walls. Due to the laws of thermodynamics and simple evaporation, the temperature can stay below 68°F (20°C) without electricity. The clay pots are porous and allow

▲ Clay pot refrigeration with layer of sand in between.

moisture to evaporate from the surface, keeping everything cool. To build this simple device, a large clay pot is put inside a slightly larger clay pot with an inch (2.54 cm) of space between them. This space is filled with wet sand, which must be kept wet continuously. The pot is covered with a damp cloth. The brick system can be larger, with a rectangular box shape within a slightly larger rectangular brick box. The space between them is also filled with damp sand, and the whole thing is covered with a damp cloth. Vegetables can last up to 28 days longer inside this device.

5 | Gardens

In order to live off a garden, you practically have to live in it.

~ Frank McKinney Hubbard

SOIL

Yes, I am positive that one of the great curatives of our evils, our maladies, social, moral, and intellectual, would be a return to the soil, a rehabilitation of the work of the fields.

~ Charles Wagner

Soil Type

Over time, any soil can be changed and improved. This means that you could theoretically put your Zone 1 garden anywhere you want, although scoping out where the best soil is makes more sense. Armed with a little knowledge about the soil, you can grow anything. The soil is the lifeblood of your land and, therefore, you.

1. The simplest way to start making decisions about the soil is to do a pH test in the garden and orchard area. pH test kits can be found at any garden center.
2. Next, find out the drainage capacity of the soil. To do this, dig a hole 1 foot (30 cm) deep. It doesn't need to be a wide hole. Fill the hole with water, and after five minutes fill it again. See how long it takes for it to drain completely. For some people, this might be never. For others, the water may disappear immediately. If it takes more than four hours to sink in, then you have a problem.
3. Look at the plants that are already growing. Bare soil is damaged by the sun, wind, and water, which is why permaculture insists on leaving soil undisturbed. Usually a plant that has become overgrown in an area, as blackberries often do, indicates that the soil has been damaged. These plants act as pioneers and prepare the soil for the next stage.
4. Good soil maintains a balance of water, air, organic materials, and nutrients through the natural cycle of growing plants. The roots take up minerals and water out of the soil; use those nutrients to produce fruit and leaves that then drop onto the ground to decay and return back into the soil.

What is rhizobium?

Certain plants are *nitrogen fixers*. These plants live in a symbiotic relationship with unique bacteria called *rhizobium*. Almost all legumes (the pea family) and leguminous trees are nitrogen fixers, and many other types of plants, such as alder, are also nitrogen fixers. The rhizobium converts the nitrogen in the air to a chemical form and releases it into the soil where it becomes usable by plants. Without rhizobium and the nitrogen fixers, nothing on earth would grow.

Planting Seeds

Seeds germinate when the soil is within a certain temperature range. Usually this is between 50–70°F (15–20°C). So, when we are waiting for spring planting, what we are really waiting for is the soil to warm up to 50°F (10°C). We can warm the soil with plastic to speed up the process. Some seeds, like apples, wild rice and some berries, require cold temperatures to sprout. They should be stored in the refrigerator all winter, with the wild rice, also kept in water. When they begin to sprout, they can be planted outside.

Some seeds also need light to germinate. Carrots, lettuce, spinach, parsley, parsnips, and beets can be thrown directly on the ground rather than pushed into the soil, but they are likely to be eaten by birds. Instead, you can soak them overnight and let them sit in the light before putting them into the ground. Larger seeds are able to germinate in the dark, and some (like parsley) need the dark.

Depth is another factor that gardeners fuss over. Usually, it is recommended that you bury the seed to a depth of four times the largest diameter of the seed. The deeper you push the seed down, the darker and wetter it will be. A shallow seed depth is likely to have more light but also dry out. The seed packets will have a recommended planting depth that should be followed, and sprouting seeds indoors will be more successful than sowing directly.

The Art of Mulch

There are many ways to mulch. The method described here is called *sheet mulching* because you layer different sheets of materials together, something like a layer cake. The beauty of sheet mulching is that you can create a new garden without the hard labor of digging and turning dirt. It works for every soil type except for the hardest dirt, for which you would have to build a raised bed or follow the procedure described in the following section. Sheet mulching also stops every kind of weed, saving you even more work.

1. Plant your largest trees and shrubs. If you get this out of the way, you won't have to go back and dig through your mulch later.
2. Cover the area with a sprinkling of dolomite, and if the soil is clay, add gypsum as well. Add any nitrogen that you can get, such as composted manure fertilizer and kitchen compost.
3. If you have some spare hay that is unfit for any other use, make a 1 or 2 inch (2.5–5 cm). If the area was covered with tall weeds, you can simply cut those down and leave them lying on the ground instead.
4. Cover the whole thing with a layer of cardboard, newspaper, old drywall, non-synthetic carpet, felt underlay, or any material that is very hefty but will break down eventually. Don't allow even the smallest hole. If you do have to work around a tree, make sure it hugs the plant tightly. This layer should be 0.5 to 1 inch (1.5–2.5 cm) tall, with any non-compostable materials like staples or plastic tape removed.
5. Water it well until it is completely soaked.
6. Add 8–12 inches (20–30 cm) of old straw from a horse stable, old chicken coop sawdust, raked or old mashed up leaves, seaweed, or seagrass. These all contain vital nutrients and can be moist. According to the composting principles (see the previous chapter),

these should be brown materials rather than green, which would turn into mush and smell bad.

7. Water everything again until it is well soaked.

8. Now you can add another 1 to 2 inches (2.5 to 5 cm) of compost and manure, and another 2 inches (5 cm) of dry material like straw or leaves. Sometimes sourcing all of this organic matter can be a challenge, but there's really no wrong way to do this, so don't sweat it. The only rule of thumb is the thicker the better. If you don't have enough material, you will have to make your bed smaller rather than try to spread it thin over a wide area.

9. Plant your largest seeds, potatoes, seedlings, and small potted plants. To do this, make a hole through the mulch to the sheet material. Use an old axe or knife to cut an X in the carpet or cardboard or whatever you used as your sheet layer, and use your hands to put dirt in the hole. Stick the seeds, potato, or seedling in the soil. If you want to plant tiny seeds, sprout them first and make a line rather than an X. You don't have to push the seeds down deep. Cover them with mulch or, for seedlings, cover the base of the plant.

10. The roots won't do very well in the first year, but if you plant roots that grow down deep, they will begin to break up the soil under the mulch. The second year you can grow more of them.

11. At the end of the first summer your soil will be immensely improved. You'll need to add a small amount of fresh mulch as the season continues, and in fact annuals can tolerate food scraps from the kitchen layered under the

mulch where worms will dispose of them immediately.

12. Keep the mulch area well watered. It will take frequent watering to keep the area moist, but as time goes on, it will get better at maintaining moisture.

13. As the seasons pass, the mulch will settle and shrink. Just keep adding layers and plant new things as the old ones are eaten, adding mulch layers each time.

14. If the weeds break through, just smash them under and add wet newspaper and a layer of sawdust. Grass will give up eventually. If a strong root takes hold, dig it up, fill the hole with fresh kitchen scraps, and cover the hole with mulch. Make sure not to bury any fresh wood products. They need to be broken down in the air first. Mulch should be loose and light, with many different materials mixed together to help achieve that air texture.

Repairing very bad soil . . .

1. If the soil is very compacted, you can loosen it with a tool, but don't *turn* it. Turning the soil means flipping it over, what you often see when someone plows up a grassy field. In a big area you can plow with an implement that just loosens up the soil, such as a *chisel plow*. A chisel plow cuts a slice into the ground without turning the soil over, allowing air and water to be absorbed. The first time you do this, plow 4 inches (10 cm) deep. The second time go down to 7 inches (18 cm). On a slope, make sure to plow the channels diagonally instead of straight downhill, which will prevent water from flowing away too easily. In the smaller gardens, use a garden fork rather than a plow.

2. If you planned out your land in the order outlined in this book, you have already controlled the flow of water by building *swales*. A swale is a hollow of land, which is built up to trap water flowing downhill.

Trees and plants are grown behind it to absorb the water and stop erosion.

3. The pH test that you did on your soil indicate what kinds of plants to grow in that area, or what to add to change the pH. On a pH scale the soil can either be acidic on one end or alkaline on the other. Typically, you would want the soil

Problem	Solution	Notes
Bare soil, no calcium	Calcium/silica	Add cement dust, bamboo mulch, or grain husks.
Bare soil, no nitrogen	Nitrogen/potash	Add leguminous tree mulch or manure.
Desert soil, can't retain water	Bentonite	Bentonite is volcanic clay that absorbs water.
Desert soil, too much clay	Gypsum	Gypsum allows water to penetrate the soil.
Salty soil	Raised beds	Raised beds allow salt to leech down away from plants.
Potash deficiency	Comfrey, wood ash	Potash is potassium found in organic matter best added by wood ash or comfrey.
No trace elements	Mulch/compost	Organic compost contains trace elements and bacteria plants need.
Too alkaline (low pH)	Sulfur	High alkalinity prevents plants from using soil nutrients. Increase to 6.0–7.5pH.
Too acid (high pH)	Lime (calcium carbonate)	Decrease to 6.0–7.5 pH

to be as *neutral* (in the middle) as possible. Neutral means a pH of around 7.

4. Any soil, not just bad soil, should be improved by planting cover and green manure crops or adding composted animal manure. You can also add *compost* (kitchen and yard waste which has decomposed) to the small gardens near the house. Soil that has been cleared most likely needs extra help, because minerals have been leeching out.

5. Encourage worms and other beneficial creatures to live and grow in the soil. These are the best cultivators, which do their own composting and mulching.

6. Land that has experienced extreme erosion needs gentle treatment. Avoid grazing any animals on it for a long while, and it may be a good idea to plant a crop of plants that grow deep roots, such as daikon radish, chicory, and leguminous trees.

7. Deep-rooted trees pull nutrients from the deepest layers of the soil, and their leaves can be used as mulch to return those nutrients back to the soil. The goal is to cover all exposed soil with quick-growing local species of trees and shrubs to prevent erosion.

Hugelkultur

This is just a fancy word that describes a method of burying old wood in the ground under a garden bed. It is a very old strategy that patterns the exact processes of the forest and lends itself very well to land that has been cleared of trees to be used for farming. Hugelkultur returns massive amount of organic material to the soil, retains huge quantities of water for the plants growing in it, and over time breaks down, leaving air pockets that provide necessary oxygen to plants.

Certain types of wood may not be suitable. Treated wood, cedar, and black locust don't rot very well, and woods that are naturally toxic to plants like black walnut are probably not conducive to growing a garden. You can use fresh wood, but it will use nitrogen to decompose, robbing it from the soil and locking it up. Rotten wood won't do this, and in fact if it is very well rotted, it may release nitrogen. Wood that has been laying around for a couple of years is probably the best material, and this includes any kind of wood material like brush, tree stumps and roots, and general debris.

The most difficult part of this is moving the wood and arranging it into the shape that you want. It's much easier and faster to use a tractor. If you are working with straight pole logs, a straight bed is easier to make, but if you are working with brush, then you can make curved beds. The wood doesn't have to be very high, but over time, the whole thing will settle as the air pockets fill with soil and the organic material breaks down. Because of this, the taller your bed is in the beginning the better. The wood can be a couple of feet tall (almost 0.5 meters), and the soil and mulch can be a couple of feet more. Some have made beds 6 feet tall (1.8 meters).

▾ Hugelkultur beds start out very tall and settle over time.

Strategically, using fresh wood is tempting because you may be clearing ground and cleaning up, and rather than making a big ugly pile somewhere and waiting for years for it to rot, it would be much nicer to stick it in the ground to be a raised bed, especially since it still absorbs water and irrigates the bed for you. You can do this, but because wood that is decomposing robs the soil of nitrogen rather than giving it back, you must add something to the soil to break down the wood faster and add needed nitrogen. The key ingredient is urine. If you have a separating compost toilet like the one described in this book, you can simply divert your household liquids to the hugelkultur beds. Straw gathered from animal bedding can be added as a mulch layer as well. The urine helps break down the wood and provides nitrogen to the soil. The other key component is fungi, which will naturally grow and help speed the process.

Steps to building a hugelkultur bed:

1. Remove a strip of sod the size of your bed. This sod will be used as a mulch layer later.
2. Lay down your wood layer, at least a couple of feet deep.
3. Sandwich several layers of mulch and soil, including leaves, grass clippings, compost, old straw, and moldy hay. Take the layer of sod that you removed in the first step, flip it upside down and lay it down on top. Add another layer of soil over it. You can use rocks to line the side of the bed so that it doesn't erode.

4. The first time you use the bed, it is recommended to grow a cover crop such as clover. Clover fixes the nitrogen and makes it available for other plants. If you plant it in early spring, you'll have time to plant root vegetables next: potatoes, carrots, radishes, etc.

Hugelkultur lends itself readily to incorporation into two other systems. A hugelkultur bed can be used similarly to a swale, with trees planted in between beds to take advantage of the water. On a larger scale it can also be applied to chinampa systems.

Teas for the Soil

A liquefied tea is an efficient method of adding nutrients to the soil quickly. If you are adding mulch and composts to your soil anyway, then this type of tea may not be necessary, but if plants fall victim to bad weather or attacks, this can help save a crop. It is particularly useful for tomatoes and peppers.

1. Mix one part manure to three parts water.
2. Leave it to ferment for at least two weeks in a container with a loosely fitting lid.
3. Add 10–15 parts more water. It should look like weak tea.
4. As you use it, keep adding more manure and water so you can have a continuous supply.
5. Extras can be added to the tea, including comfrey (which adds potassium), seaweed, or kitchen compost.

MIMICKING FORESTS

If a man walks in the woods for love of them half of each day, he is in danger of being regarded as a loafer. But if he spends his days as a speculator, shearing off those woods and making the earth bald before her time, he is deemed an industrious and enterprising citizen.

~ Henry David Thoreau

The forest ecosystem strategy:

In all natural ecosystems, plants are of different heights. Have you ever been hiking in the wilderness and come across a place where all the plants are the same size, growing in perfect rows? That kind of design doesn't exist in nature, and so we want to mimic the forest when we plant our gardens. Big trees form a canopy over smaller trees, which cover shade-loving shrubs, which in turn shade groundcover herbs. On the edges of this forest you will grow edible plants that like the sun. These form a self-sufficient community that maintains itself with less effort on your part.

The characteristics of trees are the main reason why this method is so successful. A tree is a huge biomass that affects everything around it. By its sheer size it provides homes for many creatures and insects, all of which also use it for food. These creatures often distribute the seeds of the tree in return. The roots have fungi that benefit the soil, and trunks and leaves provide shelter from the wind. Even more importantly, the tree changes the temperature and climate around itself. A large oak tree can *transpire* (or release through evaporation) 40,000 gallons of water per year. Not only is this critical for the earth's water cycle, but it also cools the surrounding air and helps create precipitation. On top of all of this, the

structures of the tree store water in the canopy and bark, and from there water runs off down to the plants and soil below.

1. To mimic this natural pattern, pick plants to grow together based on height, climbing ability, tolerance to shade, and water requirements.

2. In a dry or a cool climate, space the plants farther apart so light and water can be distributed better. In a warm, humid climate put them closer together, although not too close or you may have fungus problems.

3. In a very fertile place with lots of water, put in all of the plants at once. For example, you would plant walnuts (which live a long time), fruit trees like peaches (which don't live as long), legumes like autumn olive (as a nutritious mulch), and perennials like comfrey and yarrow (for weed control).

4. In general, you will not get as much food or useful things from each species as you would if you were farming traditionally. However, if you add up all of the *yields* (produce) from all of your crops, it will be more than what a traditional farm will have. *Diversity*, or many species living together, also gives you more security. If you have a bad year for vegetables, then you still have tree fruits, nuts, or other things.

5. Normally, in many gardens and farms, crops would be *rotated*, or moved around, to prevent pests or disease. The ground might be left *fallow*, or unproductive for a season, to give the soil a rest. When planting all kinds of different crops in one place and allowing them to mimic a forest, rotation is unnecessary. Animals are placed there for a time, and plants are started at different times, but the pattern follows the natural plant cycles.

6. However, massive amounts of plants and animals aren't added at random. To create an orderly system that works for you rather than against you, you have to create as many relationships between the plants and animals as you can (the same way you did with the elements when you designed your farm).

Establishing the system:

In a regular garden, everything is kept at the Stage 1 or pioneer level, with

herbs, weeds, vegetables, or grass growing continuously. Many people use up a great deal of energy by weeding, fertilizing and turning the soil to keep it perpetually in this stage. The opposite is true in nature. Left to themselves, plants will grow (pioneers, herbs and shrubs, and trees) that prepare the way for the next stage. Stage 1 fixes the nitrogen and breaks up the soil so that the later stages can flourish without any effort by humans. By using and accelerating this process, you can use less energy and get more food.

It is important to note, however, that the end goal is not to destroy everything that is already there. We don't want to destroy the landscape or replace all of the existing plants. When you assessed and evaluated your land, you took an inventory of all the major landmarks and plants that were growing. If you have decided to establish your garden in a place where something else already is, then hopefully these two elements will have a mutually beneficial relationship. Our "improvements" should become part of the landscape, not a replacement.

1. When you start developing a piece of land, inevitably a pioneer layer will already be growing. These so-called *weeds* can be used to build the soil. First, cut down any woody weeds so that they are lying on the ground. These can become part of the mulch.

2. If you want to speed up the process, dig out any large perennial roots. Don't dig too much or you'll encourage more weeds to grow.

3. Lay down cardboard or old carpet on top to mulch and decompose the weeds. There are two kinds of mulch, dead and living. Cardboard is dead mulch, just like leaves, straw, or dead plants, and living mulch is made of the small plants that grow under shrubs and trees. Dead mulch has to be collected and carried in, while living mulch has to be cared for over time, so each has its drawbacks.

4. When the weeds have been broken down and the soil seems ready for planting, fence off the area and begin growing legumes and shrubs that grow well in your local climate and that are useful to you, such as comfrey.

5. The soil will probably not be in the most optimum condition for planting less hardy species. Add mulch, green manure crops, and compost to improve the soil. Allow geese and ducks in to forage.

6. Plant a group of trees as a sort of *nucleus* or center of your garden. Chickens may sometimes be allowed in to forage at this phase.

7. Once all of these plants have become well established, it takes simple but careful management to make the garden sustain itself. Pigs and other animals may be allowed to forage, the trees produce fruit and mulch, and the smaller plants can be harvested on an ongoing basis.

8. After you harvest the main crops, it is a good idea to plant a cover crop or green manure to protect the soil. This usually happens in the winter. Rye,

clover, buckwheat, barley, oats, and vetch are all common cover crops, which you can harvest. Leguminous green manures like clover, vetch, and field peas can be plowed into the soil or mulched before they flower (when the beneficial nitrogen is used up).

Choosing the right plants:

Choosing the right species of plants for a specific job is probably the most important task in the permaculture design process. When choosing a species, we have to ask:

- What stage is this plant from?
- Is it a pioneer plant, or a Stage 2 or 3 plant?
- Is it deciduous (lose its leaves in the fall) or evergreen?
- How high does it grow?
- Do its roots invade the space of other plants?
- How quickly does it grow and die?
- Are the leaves dense and shade the ground, or do they allow light in?
- Is it disease resistant?
- Is it sensitive to pruning?
- Will it work in my climate and soil?
- Will it spread quickly and choke out other plants?
- Is it common or rare?
- Is it useful to me?
- Will it grow too big for the space I have for it?
- Will it take too much work for what I will get out of it?
- Can I pick species that produce at different times of the year for more yield?
- Will it produce something for a longer period of time?
- Will it produce more than one product, like leaves, roots, seeds, or fruit?
- Will it store itself, like nuts or roots that are harvested when I need them?

Just like the inventory of elements on cards you made in the beginning, species should be recorded and indexed on 3 x 5 cards. All of the above answers should be noted, as well as the general growing instructions, such as type of soil or amount of sunlight the species likes. You should also write down what it is specifically used for, such as the type of food, the animal that eats it, the type of nutrient it puts into the soil, or what color dye it makes.

Can I ever plant a big crop?

After reading about the forest garden strategy, you are probably wondering if you can ever plant a large crop of plants instead of small bits of things here and there. The answer is yes, if it's the right crop. The motivation to plant a large crop usually comes from the desire to make a profit on something that is worth some money or is in high demand. It may also be something that you eat a great deal of, like potatoes or fruit. You should only do this if the crop doesn't need much work after it is planted and is easy to harvest and store. These types of crops should never be grown on a large scale at the expense of any of the other elements.

You may also be wondering how a forest garden makes any sense. Regular monoculture farming has proven itself to be productive and efficient, allowing plants to grow without competition. They can grow as large as possible and produce the maximum amount of food per plant. Success is measured solely in pounds of food collected, and it can be impressive. However, monoculture is concerned with the individual plant and how to push each one to the greatest yield. Polyculture, and in particular a forest garden, is concerned with the whole. Pushing as many species as possible to work together actually yields more in total than one species grown alone. The plants aren't as big as possible and may not produce as much fruit, and there may be so-called "weeds," but when it is all collected in the end, the quantity of food will be greater, with the additional benefit of being self-reliant. This means more food for less work.

A word of warning, however: Forest gardens are experimental and unique. They take at least a decade to establish and even longer to make profitable. There is also no strict formula that can be followed, and every climate and microclimate is different. Cash crops with this method are a challenge, but the goal is to have many eggs in many baskets, rather than one crop that can fail. Furthermore, forest gardens are intended to require very little labor, which is the limiting factor on most market farms. Running a successful organic market farm takes 80 hours per week or more by many individuals, while a forest garden should need only a few people working part time.

PUTTING THE PLANTS TOGETHER

My green thumb came only as a result of the mistakes I made while learning to see things from the plant's point of view.

~ H. Fred Dale

Plant Communities

Once you are aware of the forest garden strategy and you know what plants you want, you can begin arranging them into communities, sometimes called *guilds*.

▾ The "Three Sisters": corn, beans, squash

These communities are a way of organizing plants around a central element based on *companion planting* (choosing plants that grow best together) and growing tendencies. This reduces competition from roots, provides shelter from the elements, adds nutrients to the surrounding soil, and deters pests.

Companion planting has been used for thousands of years, with the best known being the Three Sisters that was commonly planted by the native people of the Americas. The "Three Sisters" are corn, beans, and squash. The corn provides a support for the beans, and the squash shades the ground, preventing weeds from growing. Together they produce much more food per square foot than they could when spread out on their own. Another example is the apple tree community. Several rings of plants are grown under the apple tree's canopy. At the base of the tree is a legume like fava beans, followed by a groundcover of clover with dandelions and chicory mixed in, surrounded by a ring of comfrey, artichoke, yarrow, nasturtiums, and dill, and enclosed by daffodils that stop the grass from encroaching. This community is more complex than the Three Sisters, but each plant serves a purpose: attracting bees, providing mulch, fixing nitrogen, and hindering the growth of grass cover.

Sometimes, some of these companion plants must be grown in succession to take full benefit of one another. For example, the corn must be planted before the beans and squash, or the beans will grow too quickly and knock the corn over. Pay attention to the growing times and use your common sense when planning your garden. Observation is key in this process; keep track over time what plants grow well together and even what plants have sprung up on their own in close proximity to others. Even more apparent may be the plants that *don't* do well together. The success of a plant community often has less to do with some unseen force and is likely to come from a directly observational characteristic (like the ones you wrote down in the beginning when arranging your elements). This includes attracting or repelling deer or birds, attracting or repelling different insect species, or the function of the leaves and whether they shade the sun or mulch the ground. The one invisible trait that you should also research is the roots. The plant may fix nitrogen, loosen the soil, or provide a buffer that protects other plants from natural herbicidal plants.

Plants can be placed strategically to:

- **Attract predators:** These plants provide food or shelter to friendly insects that eat pests.
- **Sacrifice themselves:** These plants attract pests, and the pests leave other plants alone.
- **Trap pests:** These plants attract and kill pests or trap them so you can do it. Care must be taken to avoid providing a hotel for pests. Some plants provide a nice home for pests to live in all winter, giving them even more opportunity to destroy your crop in the summer.

- **Provide nutrients:** These plants can be grown to fix nitrogen or create other nutrients and friendly bacteria. They can also be cut down and left as a mulch below trees or between crops.
- **Create shelter:** These plants prevent frost, stop the wind, make mulch, and create microclimates.

Animals can also provide strategic benefits:

- **Foraging:** Foraging is a benefit. When fruit falls on the ground, flies and other pests gather. Pigs and birds can clean these up, simultaneously fertilizing as they go.
- **Insect control:** Birds that eat larvae and eggs from the bark of trees can be attracted with flowers and herbs. The flowers and herbs attract insects, which in turn attract birds.
- **Slug and snail control:** Ducks can be allowed into the garden from fall to spring, where they will totally control the slug and snail population. They stay in the marsh in the summer.
- **Predator control:** Foxes, deer, and rabbits can all be controlled with a couple of dogs. If the dog is raised with chickens, it will leave the poultry alone and protect them from predators.

How do I keep them all organized?

It can be difficult to plan communities of plants and animals when you are working with so much diversity. How do you make sure everything in a group works together, without losing your mind? This is done through a *coaction study*. Coaction means "acting together," and so when you do a coaction study, you are finding out which plants act together for the most benefit. It's actually quite difficult to go

▲ A house for carpenter bees—gentle bees that are fantastic at pollinating.

wrong in this process, since most living things will grow together without any affect on each other at all. Your goal is to try to find the ones that help each other. This is important when you are working with many communities that all interact with each other in one mega-community.

1. When Species A and Species B are put together, they can affect each other in one of three ways: positively, negatively, or not at all.

2. Obviously, if both species were negatively affected, you wouldn't want to put them together in the future. Things are not so clear if one is positively affected and the other is negatively affected, or not at all. If Species A is benefited but Species B is negatively affected, Species A is a parasite.

3. Begin observing each species in a community to see how it is doing in relation to the species around it. For

	Tomato	Carrot	Lettuce	Radish
Cucumber	-	+	0	+

example, we know from the age-old knowledge of companion planting that these plants will affect cucumbers:

- = negative
+ = positive
0 = none

4. We could also observe our local community to find out what plants grow well together in our area. If we wanted to see if the companion planting rules hold true, we could observe gardens to find out how the cucumbers are doing and tally the results. The number of cucumber plants that do well can

	Tomato	Carrot	Lettuce	Radish
+	5	6	7	8
-	5	3		1
0		3	3	1

be written rather than a + or -. For example:

5. When the results are compared, they can be slightly different than expected. Cucumbers aren't supposed to do well with tomatoes, and yet the results are 50/50 positive and negative. Lettuce is supposed to be neutral, and yet cucumbers seem to be growing well with it.

6. Once we have studied and researched these species and how they do within a group, we can begin to put them together so that each one has a +. Every plant should benefit from something, and it should be removed from any species that affects it negatively.

Companion Plant Table

Key:

Nightshade = tomato, potato, pepper, etc.

Brassica = cabbage, broccoli, kohlrabi, brussels sprouts, cauliflower, etc.

Fruit tree = apple, peach, pear, etc.

▲ Calendula

Common name	Latin name	Does well with	Does poorly with	Repels	Attracts
Alfalfa	Medicago sativa	Cotton			Assassin bug Big-eyed bug Ladybug Parasitic wasp
Amaranth	Amaranthus L.	Sweet corn			
Anise		Brassica family Clover Coriander (improves growth and flavor)			
Asparagus	Asparagus officinalis	Comfrey Coriander Dill Marigold Nasturtium Parsley and basil (deters Asparagus beetle) Tomato	Garlic Gladiolus Onion family Potato		
Basil	Ociumum basilicum	Anise Asparagus Chamomile Everything (improves flavor and growth) Nasturtium Oregano Pepper Tomato	Rue	Asparagus beetle Mosquito Fly	Butterfly

Common name	Latin name	Does well with	Does poorly with	Repels	Attracts
Bean family	Phaseolus	Beet Brassica family Carrot Corn Cucumber Eggplant Grain Lettuce Mustard Pea (improves growth) Potato Radish Rosemary Savory Spinach Strawberry	Brassica family Marigold Nightshade Onion family Wormwood	California beetle	
Beet	Beta vulgaris	Bean Brassica family Cabbage Catnip Celery Cucumber Lettuce Mint family Onion family Radish	Pole bean (stunts growth)		

Common name	Latin name	Does well with	Does poorly with	Repels	Attracts
Borage	Borago	Brassica family Cucumber Everything (improves flavor and growth) Tomato Squash Strawberry		Pest	Predatory insect Honeybee
Brassica family	Brassica	Borage Celery Chard Chamomile Cucumber Dill Geranium (traps cabbage worms) Hyssop Mint Nasturtium Onion family Pea (improves growth) Rosemary (repels cabbage fly) Sage (deters pests and improves growth) Spinach Tansy (deters cutworm and cabbage worm) Thyme	Dill Mustard Nightshade Pole bean Strawberry		

Common name	Latin name	Does well with	Does poorly with	Repels	Attracts
Buckwheat	Fagopyrum esculentum	Anything			Ladybug Tachinid fly Minute pirate bug Hoverfly Lacewing
Caraway	Carum carvi	Strawberry			Parasitic wasp Parasitic fly
Carrot	Daucus carota	Bean Chervil (deters Japanese beetle) Flax (protects roots) Lettuce Onion family Parsley Pea (improves growth) Radish (deters cucumber beetle, rust fly, and disease) Rosemary Sage (deters rust or carrot fly and improves growth)	Dill Parsnip Radish		Assassin bug Lacewing Parasitic wasp Yellow jacket
Celery	Apium graveolens	Bean Brassica family Daisy Onion family Spinach Tomato	Corn Aster flower		

Common name	Latin name	Does well with	Does poorly with	Repels	Attracts
Chamomile	Matricaria recutita	Basil Cabbage Cucumber Herbs (increases oil production) Onion family			Hoverfly Wasp
Chard	Beta vulgaris	Cabbage			
Chervil	Anthriscus cerefolium	Brassica family Carrot Grape Lettuce Rose Tomato		Aphid	
Chive	Allium	Apple Brassica family Carrot Grape Rose Tomato	Bean family Parsley Pea	Aphid Cabbage worm Carrot fly	
Cilantro	Coriandrum sativum	Bean Pea Spinach		Aphid Potato beetle Spider mite	Tachinid fly
Clover	Trifolium	Apple			
Collard	Brassica oleracea		Tansy		

Common name	Latin name	Does well with	Does poorly with	Repels	Attracts
Coriander	Coriandrium sativum	Anise			
Corn	Zea mays	Amaranth Bean Cucumber Onion Parsley Pea (improves growth) Potato Pumpkin Radish Squash Sunflower	Celery Tomato		
Cucumber	Cucumis sativus	Bean Beet Brassica family Carrot Corn Dill Lettuce Marigold Nasturtium Pea (improves growth) Radish (deters cucumber beetle, rust fly, and disease) Sunflower	Potato Sage Tomato		Ground beetle

Common name	Latin name	Does well with	Does poorly with	Repels	Attracts
Dill	Anethum graveolens	Brassica family Corn Cucumber Lettuce Onion family	Cabbage Carrot Tomato	Aphid Cabbage looper Spider mite Squash bug	Honeybee Hoverfly Ichneumonid wasp Tiger swallowtail butterfly Tomato horn worm Wasp
Eggplant	Solanum melongena	Bean Four-o'clock Marigold (deters nematode) Mint Pepper Spinach Tarragon	Pepper Potato		
Fennel	Foeniculum vulgare	Dill	Everything	Aphid	Ladybug Syrphid fly Tachinid fly
Flax	Linum usitatissimum	Carrot Potato			

Common name	Latin name	Does well with	Does poorly with	Repels	Attracts
Fruit tree		Clover Comfrey Daffodil Leek Nasturtium Onion family Parsnip (attracts predatory insect)	Cedar Nightshade family Pepper Walnut		
Garlic	Allium	Celery Cucumber Everything (deters aphid and beetle) Fruit tree Lettuce Rose	Asparagus Bean Parsley Pea	Ants Aphid Cabbage maggot Cabbage worm Carrot fly Rabbit Slug	
Grape	Vitis	Chervil (deters Japanese beetle) Chive Hyssop Nasturtium Radish Tansy			
Hemp	Cannabis sativa	Brassica family Grape		Cabbage moth larvae Cabbage butterfly	Butterfly Honeybee

Common name	Latin name	Does well with	Does poorly with	Repels	Attracts
Hyssop	Hyssopus	Cabbage Grape	Radish		
Lettuce	Lactuca sativa	Bean Beet Carrot Chervil Cucumber Dill Kholrabi Mint (repels slug) Onion family Pea Radish Strawberry	Cabbage Chrysanthemum Cress Parsley		
Lovage	Levisticum	Everything			Ichneumonid wasp Beneficial ground beetle
Marigold	Asteraceae calendula	Asparagus Brassica family Cucumber Eggplant Pepper Tomato		Beet leaf hopper Beetle Nematode	

Common name	Latin name	Does well with	Does poorly with	Repels	Attracts
Mint family	Mentha	Beet Brassica family Eggplant Lettuce Strawberry		Ant Cabbage fly Cabbage looper	
Mustard	Brassicaceae	Bean Brassica family Turnip			
Nasturtium	Tropaeolum majus	Apple Bean Brassica family Cucumber Everything Tomato		Aphid Cabbage looper Cucumber beetle Squash bug	Predatory insect
Onion	Allium	Beet Brassica family Carrot Fruit tree Lettuce Nightshade Carrot (together deter rust fly and nematode) Dill Celery Cucumber Squash Strawberry Fruit tree	Asparagus Bean Pea Parsley	Slug Aphid Carrot fly Cabbage worm	

Common name	Latin name	Does well with	Does poorly with	Repels	Attracts
Oregano	Origanum vugare	Everything		Aphid	
Parsley	Petroselinum crispum	Asparagus Carrot Corn Rose Tomato	Onion family		
Pea	Pisum sativum	Bean family Cabbage Carrot Corn Cucumber Lettuce Pea Potato Radish Spinach Squash Turnip	Garlic Onion family Potato		

Common name	Latin name	Does well with	Does poorly with	Repels	Attracts
Pepper	Solanaceae	Basil (provides ground cover) Carrot Eggplant Four-o'clock Marjoram Mint family Onion family Tomato	Bean Black walnut Brassica family Corn Dill Fennel		
Pole bean		Carrot Corn Cucumber Eggplant Lettuce Marigold Peas Radish	Bean Beet Cabbage Garlic Onion		
Potato	Solanaum tuberosum	Bean Cabbage Corn Flax Horseradish or tansy (deters Colorado potato beetle) Marigold Mint family Onion family Pea (improves growth)	Brassica family Carrot Cucumber Eggplant Fennel Pepper Pumpkin Raspberry Squash Sunflower Tomato		

The Ultimate Guide to Permaculture

Common name	Latin name	Does well with	Does poorly with	Repels	Attracts
Pumpkin	Curcurbita	Bean Buckwheat Catnip Corn Radish (repels flea beetle) Tansy	Potato	Spider Ground beetle	
Radish	Raphanus sativus	Bean Beet Carrot Chervil Corn Cucumber Grape Lettuce Pea (improves growth) Spinach Squash	Hyssop		
Rose	Rosaceae	Chervil (deters Japanese beetle) Chive Garlic Parsley			
Rosemary	Rosmarinus	Bean Cabbage Carrot Sage		Bean beetle	

Common name	Latin name	Does well with	Does poorly with	Repels	Attracts
Rutabaga		Pea (improves growth)			
Turnip					
Sage	Salvia	Bean Beet Brassica family Carrot Cauliflower Rosemary Strawberry Tomato	Cucumber Rue	Black flea beetle Cabbage fly Cabbage looper Cabbage maggot Carrot fly	Cabbage butterfly Honeybee
Savory	Satureja	Bean Brassica family Onion family			
Spinach	Spinacia oleracea	Bean Cabbage Celery Eggplant Pea Radish Strawberry Tomato			
Squash	Cucurbita	Borage Corn Nasturtium Onion family Radish (deters cucumber beetle, rust fly, and disease) Tansy			

Common name	Latin name	Does well with	Does poorly with	Repels	Attracts
Strawberry	Fragaria ananassa	Bean Borage or sage (enhances flavor, deter rust fly, and disease) Lettuce Mint family (deters aphid and ant) Onion family Spinach	Brassica family		
Sunflower	Helianthus annuus	Corn Cucumber Tomato	Potato	Aphid	
Tansy	Tanacetum vulgare	Bean Cabbage Corn Cucumber Grape Potato Rose Squash	Collard	Ant Flying insect Japanese beetle Squash bug Striped cucumber beetle	Honeybee
Tarragon	Artemisia dracunulus	Everything			
Thyme	Thymus	Borage Brassica family Tomato (with cabbage deters flea beetle, cabbage maggot, cabbage butterfly, colorado potato beetle, and cabbage worm)			

Common name	Latin name	Does well with	Does poorly with	Repels	Attracts
Tomato	Solanum lycopersicum	Asparagus Basil Borage Carrot (stunts carrot growth but grows better) Celery Chervil (deters Japanese beetle) Garlic Geranium Marigold Nasturtium Onion family Oregano Parsley Pepper Rose Sage Spinach Thyme	Bean Black walnut (inhibits growth) Brassica family Corn Dill (attracts tomato horn worm) Fennel Pea Rosemary	Asparagus beetle	

DESIGN

If design were reducible to a set of principles, wouldn't we find an awful lot of similar houses, gardens, cars, rooms? You'd have no variety.

~ Gitte Lindgaard

Trellises

The house, the fence, the outbuildings, the shaderoom, the patio—these are all possible *trellises*. A trellis is simply a framework that supports plants. Trellises can be built over pathways and in the middle of gardens in the form of tripods,

▲ Wire, wood, and rope trellising.

circular baskets, wire frames, archways, or even over streams. They can create hedges by planting perennials on them (such as hops), they keep areas cool, and they become part of living spaces. Ivy, grape, wisteria, roses, beans, cucumbers, melons, squash, peas, and tomatoes all need trellising.

City Gardens

In a suburban area you will probably only have room for Zone 1 and maybe Zone 2 if you are lucky. This means that you will have to grow things very intensely. Even those who don't have a yard and are working with just a patio or windowsills can grow a surprising amount of food in pots. Any container can work, as long as you cut a few drainage holes in the bottom.

Root crops like potatoes need to be grown in a deep metal drum, wood box,

or stacked tires, or in a bed of mulch and covered with more mulch. As they grow, put more mulch around the stem so that it is forced to grow up out of the barrel. As you fill the barrel with mulch, the stem will grow higher to reach towards the light, and this will force it to produce more potatoes.

Super nutritious varieties that work particularly well for container gardening are:

- Pepper
- Parsley
- Tomato
- Chive
- Swiss chard
- Lettuce

If you have very little space, stick with herbs. Another way to maximize your space is to build upwards. Hang baskets, build a greenhouse that sticks out of the

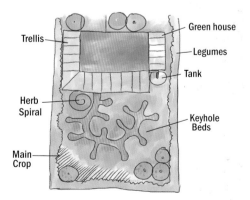

▲ This city garden uses trellises as fencing.

window, and build shelving up the walls. Of course, it's always a good idea to be growing sprouts in the kitchen and a sack of mushrooms in a dark place.

If you do have room for a fruit tree, it should be a dwarf variety. If you already have a fruit tree, keep it pruned back so that it doesn't shade your garden. When you plant a new tree, you can plant it against a wall or fence and train it to grow flatter. It will have to be tied and pruned, but it will save an amazing amount of space.

There are tremendous resources available in the city because of your close proximity to people. You will have to bring in organic material, but you should be able to find it where it would otherwise go to waste. City maintenance for the parks, construction projects, neighbors, and Internet communities like Freecycle can all offer sources of compost, wood brush, grass

clippings, and other valuable "garbage" that you can use. Your neighbors are likely to be growing plants too, and bartering can offer more variety to your diet.

One concern with city food production is lead. Lead can come from paint that was used on a house or even dumped by some irresponsible person decades ago, and it can come from the exhaust of nearby cars on a busy road. If you suspect that there may be lead in the soil, get it tested. The solution is to add as much organic material to the soil as possible, which will make the lead less available and thus less likely to be absorbed by a plant. Be aware that if you gather organic materials from the city, you should avoid collecting leaves that were right next to a busy highway. Choose leaves that are on quiet side streets.

Desert Gardens

The desert provides unique challenges and opportunities because of the heat and shortage of water. There are three

▾ Edible prickly pear cactus.

strategies that will contribute to your success:

- Create successions that are self-shaded.
- Use edible tree species that thrive in dry climates as your staples.
- Use boxes and trellises as close to the house as possible or attached to it.

Your *staple* foods are olives, palms, citrus, avocado, apricot, banana, and maybe papaya. Staples are foods that provide you with half of your diet when they are in season because they are nutritious and produce lots of food. The graywater marsh and other rainwater runoffs feed into swales directly to the trees, which also simultaneously shelter the house and create a cooler climate. Well-mulched raised beds made of mud, with the shade of wood trellises, vines, or trees, are planted with peas and fava beans in succession with celery, onion, carrots, spinach, tomatoes, peppers, and melons. Climbing

vines like grapes can be grown year round in succession as well.

GROWING

Nature gives to every time and season some beauties of its own; and from morning to night, as from the cradle to the grave, it is but a succession of changes so gentle and easy that we can scarcely mark their progress.

~ Charles Dickens

Care and Succession

Many people plant a garden, but once the seed or seedling is in the soil, they don't know what to do other than give it water and hope for the best. However, there's still more work to do. The next step for vegetables is to get the plant to flower, because without flowers there aren't any vegetables. The next

▾ Transplants are the simplest way to extend the growing season.

step for herbs and lettuces is to prevent flowering, because otherwise they will stop producing leaves.

As the season progresses, you will also need to replant. Eventually, no matter what you do, the lettuce will *bolt* and stop making more lettuce leaves for you. With careful planning, you can also extend your growing season far beyond what it otherwise would have been. In the following schedule, the seeds are planted together randomly so that all of the species are mixed together.

Before the first frost: Around two weeks before the first frost of the season, you can start sprouting some quick-growing cold hardy brassicas inside. This includes cabbage and broccoli.

Last frost date: Once you know that the last frost has passed, plant cold hardy plants outside. This includes lettuce, kale, dill, radish, parsnip, mustard, arugula, and carrots. Plant these close together, especially the lettuce.

One month after last frost: The radishes should be ready to eat, and the brassica seedlings can be transplanted into the space where the radish was. The other herbs and leaf vegetables should begin to be ready as well.

Early summer: When the soil has warmed up, remove a few whole heads of lettuce and plant bush beans in the spaces where they were. The cabbages will begin ripening, and the remaining greens will try to bolt, so make sure you harvest those leaves quickly to prevent this.

Fall: You can finally harvest parsnips, and as the plants begin to be depleted, you can put in fava beans or push garlic cloves into the ground.

You will notice that there is a great deal of crowding going on. Rather than waiting for lettuce to get big, you are removing the whole lettuce as it begins to crowd everything around it. Don't remove all of the lettuce heads but just enough to give other plants room. Then continue to harvest the leaves from the other plants.

Replanting Annuals the Easy Way

These techniques work well only in a mild climate, but it's possible you could use cold frames or the greenhouse to try to do this as well. Certain plants can be made perennial:

Leek: Allow some leeks to go to seed. At the end of the season you can dig them up where you will find small secondary bulbs growing off the base of the stem. These can be planted just like onions. Alternatively, you can cut off the leek at the ground at harvest instead of pulling it out, and it will grow a second time.

Garlic: Garlic can be a perennial. If you leave it in the ground for a couple of years, it will begin giving you an everlasting crop of garlic.

Broad bean: Large pods often grow near the base of a broad bean plant, where they can be left on the ground to dry. In late summer, mulch over the top of them with straw, and they will sprout in the fall.

Potato: In the fall, leave seed potatoes in the ground and mulch them well. In the spring they will sprout.

Lettuce: Allow lettuce to go to seed, and it will scatter seed and replant itself.

Fruit and melon: Tomatoes, pumpkins, and other melons can be left

▲ A recycled water bottle acts as a cloche, or mini-greenhouse.

in the garden and covered with mulch at harvest time, where they will rot and spill their seeds. These will grow and effectively replant themselves.

Carrot: When you eat carrots, keep the tops and store them in a dark cool place. They will begin to sprout, and you can set them out to grow.

Cabbage: Cut the stalk high on the plant, leaving a few leaves. Little cabbage heads will spring up out of the stalk, and they can be eaten or replanted.

▲ Cold frames can be covered with glass or plastic.

These plants are cold-hardy and grow quickly
Anise
Basil
Broccoli
Caraway
Celery
Chamomile
Cherry tomato
Chervil
Chive
Cilantro
Cucumber
Endive
Japanese greens
Lettuce
Mint
Nasturtium
Okra
Oriental cabbage
Parsley
Rosemary
Sage
Swiss chard
Tarragon
Thyme
Turnip tops

Cold Climate Gardens

In a colder, temperate climate you probably won't be able to grow enough food all year to feed an entire family. Realistically, that's very difficult. You will be able to supplement your supply with the greenhouse, but you'll have to strategize the rest. We do this in two ways: extending the growing season and preserving food. To extend the growing season, we create a rotating planting season. We plant cold hardy salad vegetables in early spring, then our summer vegetables, followed by more cold hardy salad plants, then root crops and vegetables that can stay in the ground during the winter. In some places we can also grow a green manure crop before spring.

When there is threat of frost, some of these plants must be protected by various devices. A *cloche* is a fancy word for a miniature greenhouse. It can be built as a hoop house over a raised bed covered with plastic, or as a tiny glass box that is placed over a bed. Many people also use plastic milk jugs as a cover. Cut the bottom off the jug and stick it over the plant you want to protect from frost. These devices are simply popped over the plants at night and lifted off first thing in the morning. Another variation is the *cold frame*, which is a wooden box, usually a raised bed with a glass lid that opens on a hinge. Old house windows work well for the lid. The window can be open during the day and shut at night.

Carrots, turnips, leeks, other root crops, and some greens like kale are hardy to frost, so you can leave them in the ground. Cover the soil with a thick layer of hay to keep it from freezing. At the end of fall you may have a bunch of unripe green tomatoes still on the vine. To ripen those last tomatoes and save them from frost, pull the whole plant up and hang it upside down in your basement or cellar.

Cold Climate Garden Schedule

April	June	August	September	October
Plant early garden transplants or seeds. Start transplants for the next planting.	Harvest early garden. Plant spring transplants. Start transplants for summer garden.	Harvest spring gardens. Plant summer transplants. Start transplants for fall garden.	Harvest spring/ summer gardens. Plant fall transplants. Start transplants for winter garden.	Harvest all gardens. Plant winter transplants.

6 | Cooking and Preserving

He who distinguishes the true savor of his food can
never be a glutton; he who does not cannot be otherwise.
~ *Henry David Thoreau*

▲ Ghee

LOW-ENERGY COOKING

Everyone is kneaded out of the same dough but not baked in the same oven.

~ Yiddish Proverb

Comparing Cooking Methods

The most convenient and most ecological stove is an electric or natural gas modern appliance found in most homes. However, to become dependent on locally produced fuels, you must use wood, methane, or the sun. Wood cookstoves work the best in cold climates where firewood is plentiful. Gas stoves are better for hot climates, especially if you have your own methane digester to make gas fuel, while solar cooking is a solution that can be used in the summer on a hot day. It is important to choose the method that makes the most sense, and not for its aesthetic value.

The wood cookstoves are the most common option, but they are likely to be recycled antiques, which aren't very efficient. You can purchase brand new ones from Amish manufacturers that are better but also much more expensive. You also have to get the wood, chop it, and allow it to dry (often for six months). There's also cleaning and maintenance, which involves sticking a wire brush down the chimney and shoving the creosote and soot out into your kitchen. Not to mention that you wouldn't want to use the stove in the summer, when you need to be canning fruit and vegetables. If you rely completely on wood fuel, you will have to have an outdoor kitchen for hot weather. The other downfall of wood is the smoke; if everyone started cooking with wood, the smoke would be overwhelming, and it would be a very polluted place. On the other hand, wood is a renewable energy source and comes from the place where you would use it. You definitely need to weigh the costs and benefits. If you have your own woodlot and you build an outdoor kitchen, it's probably worth it.

Using a Wood Cookstove

One end of a wood cookstove has vents and dials, regulating the airflow and thus the heat of the fire: more air equals more heat. At the back of the firebox is a *slide unit* that directs the smoke around the oven box before going up the chimney so that the oven is heated more consistently.

1. Move the slide unit mechanism so it is directed towards the chimney. (Note that each stove is different.) This will create a draft and pull the smoke up the chimney and not into the room.
2. Start a fire with quick-burning wood that creates a lot of heat and keep feeding it to keep the temperature up. If you are good at starting a fire, it will take about

15 minutes to heat the oven to 350°F (176°C), at which point you can add a slower-burning piece of wood.

3. Once you reach a steady temperature, you can start baking and cooking. If the stove gets too hot, open the door briefly, but not for too long, and watch for hot spots on the surface. Over time you'll become familiar with the quirks of your stove.

4. The whole cooktop surface of a woodstove is hot. The warmest area is directly over the firebox, and the coolest is on the front opposite the firebox. You will most likely have to babysit the stove the whole time you are cooking, because temperatures can fluctuate quickly.

▾ Wood cookstove.

Thermal Cookers

Thermal cooking is a simple, low-energy solution that can supplement your cooktove use. A thermal cooker has two pots, one for the food and one to create a vacuum seal that insulates the food. Some thermal cookers have two small pots that fit into the outer container, so you can make two things at once. The food is brought to a boil or to cooking temperature for about 10 minutes in the small pot, and then the whole thing is placed into the outer pot and shut with the lid. It will continue to cook for six hours or more, without using any more energy.

Thermal cookers can be used in conjunction with a solar cooker. The solar cooker can bring items up to the desired temperature, but then what if it gets cloudy? Nonetheless, a solar cooker is an amazing low-energy cooking solution in sunny areas, and it is easy to make. Any cardboard or wood box can be used to fashion a solar cooker by covering it with shiny aluminum foil. The box has a large reflective shield, focusing light into the center where the food sits.

Mud Oven

The mud oven is a low-cost way of building your outdoor kitchen. It needs to be built in your shaderoom under cover,

▾ Pre-heating a mud oven.

with the chimney funneling the smoke out. There are certain proportions that have been used for centuries in mud oven design: The door is 63% of the oven's height, and the height is 60-75% of the dome's diameter. Traditionally, this means that the oven base is 27 inches (68.6 cm) wide and around 17 or 18 inches (43–45.7 cm) tall on the inside. The door is 63% of 18 inches, or 11.25 inches (28.5 cm) high. The door is half as wide as the inner diameter of the oven, so if the oven base ends up being 23 inches (58.4 cm) on the inside, the door would be about 11 inches (28 cm) wide.

The oven must be built on a stand or foundation that raises it to about waist-height to be comfortable height for baking and must be able to bear the weight of the heavy clay. A slip form rock foundation is the easiest and strongest way to do this (see the chapter on building houses), although people use poured concrete, old metal drums, or other recycled fireproof materials.

There are fancy clay oven designs that provide a separate firebox that keeps your bread away from any soot or coals, but this design is a simple, old style oven that works just as well if not better. You build up a fire inside at least a few hours in advance, let it burn down to hot coals, remove the coals and ashes, and then slide the bread in to bake.

1. Once you have built your foundation, you need to put down a layer of sand, 5 inches (2 cm) deep. Smooth the sand out and make it level.

2. Put a layer of bricks on top of the sand. It will provide insulation and also make the oven level. You can use any old bricks as long as they don't have any cement on them. When you lay them down, you must be careful not to push the sand around, and they must be snug against each other. Put them down

straight and firm without jiggling. Tap them flat so they make a level surface.

3. Now the fun part. Using wet sand or topsoil, create an oven mold on top of your bricks. The pile should be 18 inches (46 cm) tall and as wide as you can make it while still leaving room for the mud walls you'll be putting on next. Smooth your form out and make the dome nice and round. Cover it with wet newspaper.

4. It's time to mix your mud. First you need clay subsoil. This can be found by digging down past the topsoil, or you can go to construction sites, riverbanks, or other places where the ground has been cut. It is easy to recognize because it feels just like clay and is easily moldable.

5. Mix the cob or mud on a tarp with your feet: one part clay, approximately three parts sand. Mix it very well, adding a sprinkling of water now and then to keep the mixture moist. Test it by making a hard ball and dropping it on the ground. It will change shape but shouldn't crack. If it cracks, add more clay, and if it is too sticky, add more sand.

6. Start sculpting the cob mixture around your sand form without pressing into the sand. The wall will be four inches thick, so you can grab big globs and press them down gently. Some people find that using their knuckles works

▾ **Mud oven under construction with sand inside.**

better for pressing. You will be cutting the door out later so just cover the whole thing with clay.

7. Wait at least four hours to let the clay dry. Then cut the door out with a big sharp tool using the 63% rule.

8. Carefully remove the sand and newspaper from the inside. Start a very small fire inside near the door with some newspaper and small kindling. Once it gets going, push that fire to the back and start another near the door. Watch to see if the smoke pulls up inside the dome or comes out the door. If it is pulling inside, cut your door a little bigger.

9. Some people build a little brick archway with leftover mud as mortar to make the door look nicer, but either way you also need to make a wooden door. The door is made to fit and is only used when your fire has burned down to coals, to help hold in the heat.

10. You can add a little water to your leftover mud and use it as a plaster to add some extra smoothness and insulation to the outside of the oven. Don't cover it with anything waterproof. The oven has to breathe and allow steam out in order to bake anything.

Using the oven takes a little practice but is very simple. Build up a fire and leave the door off to let the smoke out. After two or three hours it will burn off the black soot on the inside of the dome. Rake out the ashes and coals. Slide in your bread and put the door on. To test if it is hot enough, throw some flour inside. It should turn brown in about 15 seconds. You can also roast, broil, steam, or braise a variety of foods, including vegetables, meats, pizza, pie, stew, casserole, cake, cookies, and beans, and when it begins to cool off you can make yogurt or dry herbs or fruit. The interior of the oven reaches temperatures of 700°F (371°C) and bakes through conduction and convection, holding the moisture in and making incredible bread. It will hold heat for several hours allowing you to make lots of food at once.

PRESERVING WHOLE FOODS

Unquiet souls. In the dark fermentation of earth, in the never idle workshop of nature, in the eternal movement, yea shall find yourselves again.

~ Matthew Arnold

Food Safety

Many people are wary of preserving their own food because of the danger of *botulism*, that scary invisible bacteria that paralyzes or even kills you. Botulism is the reason babies don't eat honey until after a year old, and why canning books tell you to boil and sterilize your canning equipment. People have gotten botulism from oil infusions left out on the counter and even potatoes wrapped in foil. Most people think they can smell bad food or see the mold, but botulism is invisible and doesn't have a taste or smell. To prevent botulism poisoning:

- Cook home-preserved food for at least 20 minutes.
- When preparing foods for preserving, wash your hands frequently.
- Keep utensils and containers extremely clean.
- Use separate utensils and containers for each type of food you preserve, especially meat.

Salting Meat

You could smoke meat instead, which tastes delicious, but smoking takes a smokehouse and several weeks of time and constant vigilance. Salting is a practical and efficient method of preserving meat with very little effort.

1. Clean the meat and cut off anything you don't like. Save the fat to make clarified fat. Dry the meat with a clean cloth and cut it into smaller strips so that it will be easier to make sure that the meat in the middle is preserved. Rub spices into the strips, and then rub tons of salt into them until you can't rub any more. There are salt curing products available made just for this purpose, but these contain sodium nitrite which you probably don't want to use.

2. When you've rubbed in as much salt as you can, cover it in a layer of salt to coat it. Hang it up in a place that stays consistently 59°F (15°C) for at least three weeks, checking often for spoilage. A basement or cold storage is ideal. The meat should stay edible for at least a few months. The way this works is that salt dissolves into the water in the meat, preventing bacteria from growing if the balance is greater than 3.5% salt to water. Ideally it should be over 10% salt. You can't really control the percentage but if you just rub so much salt into the meat that it just won't accept any more, you can be fairly sure that you've got it.

3. When you are ready to cook it, just wash off the salt well. You might have to soak it a little bit to get it all out.

Clarified Fats and Butters

Fat is useful and healthy, as long as it's used the right way. There is a big difference between the fats found in factory foods and the natural fats found in homegrown meat or dairy. You will still want to remove excess fat when you are cooking, but it serves a useful purpose.

Lard is an unpleasant word that is synonymous with clarified fat. It is fat that is cut up, liquefied, and filtered. When it cools, it becomes a block of lard. Often several types of fats are mixed together, and you can even add a small amount of vegetable oil. To make clarified fat, save all the fats from chopping up meat and store them in the freezer for when you are ready to process them. When you are ready, put them into a saucepan and simmer on low for a few hours until the fat is liquid. If you would like to make your own bullion cubes out of it, cut up onions, carrots, leeks, turnips, herbs, and spices, and add salt and pepper to the fat. Once it is liquefied, pour it through cheesecloth and allow it to cool. It should last years and can be used in soups and stews or for greasing a frying pan.

Clarified butter, or *ghee*, lasts much longer than regular butter. Once made into ghee, it can sit at room temperature for months without going bad. The butter is melted at a very low temperature until it is completely melted. Don't stir it, but you can raise the heat slightly so that it is steaming a little. Don't let it turn brown. The solids will rise to the surface and will need to be skimmed off. Eventually (usually hours later), the butter will be a golden color and completely clear. Pour it into a container and when it is solid and cool, put the lid on tightly. Clarified butter is used in cream sauces and for frying.

Egg Storage

If you have quite a few chickens, they will probably produce more eggs than you can eat. Eggs can actually last a long time on their own. In the fridge in a regular egg carton they last about six weeks, and in a plastic bag can last two months. Cold storage, pickling, drying, and freezing can extend that time and possibly get them through the winter when the chickens aren't laying.

For cold storage, pack freshly gathered eggs into a wooden, plastic, or ceramic container in sawdust or oatmeal with the small end down. Don't store the eggs near anything smelly like onions. They must be stored at 30 to 40°F (-1 to 4°C) in fairly high humidity and will last about three months.

To pickle eggs, first hard-boil them, cool immediately in cold water, and remove the shells. Put them into wide-mouthed jars. Soak them in a brine of half a cup of salt per two cups of water for two days. Pour off the brine. In a saucepan heat:

1 quart vinegar
¼ cup pickling spice
2 cloves garlic
1 tablespoon sugar

Bring this mixture to boil and pour it over the eggs. Screw the lid on tightly and leave the eggs alone for seven days to cure before eating. Pickled eggs will last four to six months in a fridge or properly chilled cold storage.

The easiest method of storing eggs is freezing. Use only fresh, clean eggs that you didn't have to clean yourself. This means eggs that happened to not get any dirt or manure on them. Crack the eggs and put their contents into a freezer container bag. Only freeze as many eggs per bag as you will use at one time because you can't refreeze the eggs once you thaw them. Stir the eggs together without whipping in any air and add:

1 tablespoon of sugar OR
½ teaspoon of salt per cup of egg

The eggs will keep for eight months in the freezer.

Drying is a somewhat more time-consuming method that extends the life of your eggs. Compared to other methods it may not be worth it since the eggs will only last three to four months, which is as long as simply putting them into cold storage. Nonetheless, you might still want to do it. Crack very fresh eggs and beat them well in a bowl. Pour them into a drying surface that is lined with plastic or foil, no more than 1/8 inch thick. Plates work for solar drying, or you can use a dehydrator. In an oven or dryer, dry at 120°F (49°C) for 24–36 hours, then turn the eggs over, remove the plastic or foil, break the eggs up and dry for 12–24 more hours. In the sun, it will take five days until they are dry enough to break easily when touched. Grind the eggs into powder and use in baking, or reconstitute by adding an equal amount of water (half a cup of egg powder to half a cup of water).

Lard is the most effective method if you have lots of lard. Use very fresh, clean eggs and dip in melted lard. Lay them out to dry, then buff them gently with a clean towel to remove any excess and to ensure that the lard is spread all over the eggs. Then pack the eggs in salt in a large bucket so that no eggs touch each other. Put the bucket in a cool place, and stored this way they will last six months to a year.

FREEZING

God comes to the hungry in the form of food.

~ Mahatma Gandhi

Blanching

Before you can freeze vegetables, you must *blanch* them. Blanching slows or stops the enzymes that make vegetables lose their flavor and color. There are two ways to do this: boiling and steaming. If you blanch too much, vegetables lose nutritional value, but blanching too little will speed up the enzyme breakdown and make them all brown and wilted. You have to stick to tried-and-true blanching times to do this right.

The day before you start, turn the freezer temperature down to -10°F (-23°C), which will help everything to freeze quickly. Be ready to label every item with the date and what it is. Don't forget to turn the temperature back to 0°F (-17°C) when everything is frozen. Frozen fruits and vegetables will last about a year, with the exception of onions, and baked foods can last six months. Animal products and meat only last three to six months.

To boil vegetables, wash them thoroughly and drain well. Some foods can be frozen whole, but check the table for which ones can. Most must be trimmed and chopped. You will need 1 gallon (3.7 liters) of water per pound (0.5 kg) of prepared vegetables, or 2 gallons (7.6 liters) per pound (0.5 kg) of leafy greens. Bring the water to a boil, and lower the food in with a wire basket, mesh bag, or metal strainer. It should take less than a minute for the water to get back up to a boil; if it takes longer, you are using too much water.

Keep them submerged for the time specified on the chart, then pull them out and quickly chill them in ice water for the same length of time that took you to boiled them. Drain them well, pack into a container or zip up in freezer bag with as little air as possible, and put them in the freezer.

Steaming is almost exactly the same, but instead of submersing foods, there is only about 2 inches (5 cm) of boiling water in the pot. The steamer basket is lowered in, and the lid is put on. You start timing as soon as the steam starts trying to push out of the lid again.

Onions, peppers, and herbs don't need to be blanched at all. Squash, pumpkins, sweet potatoes, and beets need to be fully cooked.

DRYING

Three things give us hardy strength: sleeping on hairy mattresses, breathing cold air, and eating dry food.

~ Welsh proverb

Air Drying

This method works well for many herbs as well as alliums like onions and garlic. Herbs can be tied in bunches, and onions and garlic can be braided by their tops together. They are simply hung upside down from the ceiling of a cool, dry, airy room that stays dark most of the time. Your cold storage downstairs may be too humid, but a large cool pantry or even a big cupboard will work. It should take about two weeks to completely dry, and they will hold their flavor longer if you keep them there. Just take some as you need them.

Blanching Chart

Food type	Steaming	Boiling
Artichoke	hearts: 8 minutes	hearts: 7 minutes
Asparagus	small stalk: 2 minutes medium stalk: 3 minutes large stalk: 3 minutes	medium stalk: 3 minutes
Bamboo shoots		10 minutes
Bean sprout		5 minutes
Beet greens		2 ½ minutes
Black eyed pea	2 ½ minutes	2 minutes
Broad bean pods		4 minutes
Broccoli	5 minutes	
Brussels sprout	small heads: 3 minutes medium heads: 4 minutes large heads: 5 minutes	
Butter bean	small: 2 minutes medium: 3 minutes large: 4 minutes	
Cabbage	shredded: 2 minutes wedges: 3 minutes	1 ½ minutes
Carrot	whole: 5 minutes diced/sliced: 3 ½ minutes	sliced: 3 minutes
Cauliflower	3 ½ minutes	3 minutes
Celery	diced: 3 ½ minutes	diced: 3 minutes
Chard		2 ½ minutes
Chayote	diced: 2 ½ minutes	diced: 2 minutes
Chinese cabbage		shredded: 1 ½ minutes
Collard green	3 minutes	2 ½ minutes
Corn on the cob	small ears: 7 minutes medium ears: 9 minutes large ears: 11 minutes (note: cooling time doubles)	small ears: 6 minutes medium ears: 9 minutes large ears: 10 minutes (note: cooling time doubles)
Corn cut from the cob	5 minutes	4 minutes
Dasheen	3 minutes	2 ½ minutes
Eggplant	1 ½ inch slices: 4 ½ minutes	1 ½ inch slices: 4 minutes
Green bean	3 minutes	2 ½ minutes
Green peas	2 minutes	
Greens	2 minutes	
Irish potato	4 minutes	
Jerusalem artichoke	4 minutes	
Kale		2 ½ minutes
Kohlrabi	whole: 3 minutes diced: 1 ¾ minute	diced: 1 minute

The Ultimate Guide to Permaculture

Food type	Steaming	Boiling
Lima bean	small: 2 minutes	small: 1 ½ minutes
	medium: 3 minutes	medium: 2 ½ minutes
	large: 4 minutes	large: 3 ½ minutes
Mushroom	whole: 5 minutes	medium, whole: 5 minutes
	sliced: 4 minutes	
	buttons/quarters: 3 ½ minutes	
Mustard green		2 ½ minutes
Okra	small pods: 3 minutes	small pods: 3 minutes
	medium pods: 4 minutes	medium: 4 minutes
Parsnip	3 minutes	2 minutes
Pea	edible pod: 2 minutes	edible pod: 1 ½ minutes
Pinto bean	small: 2 minutes	
	medium: 3 minutes	
	large: 4 minutes	
Rutabaga	diced: 2 ½ minutes	diced: 2 minutes
Shell bean		1 ¾ minutes
Snap beans	3 minutes	
Soybean	in pod: 3 minutes	4 minutes
Spinach		2 ½ minutes
Summer squash	2 minutes	
Sweet peppers	halves: 3 minutes	
	strips/rings: 2 minutes	
Turnip	diced: 2 ½ minutes	diced: 2 minutes
Turnip greens		2 ½ minutes
Wax beans	2 ½ minutes	3 minutes

Solar and Electric Drying

Sun drying is a no-cost, low-energy dehydration method that uses the power of the sun. This method works well in a hot, dry climate. In temperate regions a large reflector can be made to help focus the sun's rays, like a big solar cooker, but it will still take a few days if it is successful at all. The alternative is an electric dehydrator. There are many on the market today that use very little energy to power a heating element and fan, and there are some that work effectively with a fan alone.

1. Use only ripe fruits and vegetables. Wash everything thoroughly, peel and slice them very thin unless you are drying peas or corn, which can just be removed from pods and cobs. For seed pods, harvest them before they burst and use them whole. This process is the most time-consuming step. There are simple devices such as apple corers and peels that can help speed this process. Over-ripe or even fermenting fruit can still make good fruit leather. Wash, peel, remove seeds and pits, and then grind them up by mashing or blending. The puree must be thin enough to pour but not too thin to be watery. If it is too thick, add fruit juice or water; if it is too thin, add another kind of fruit puree. For meat, choose lean cuts of beef, buffalo, goat, or deer. Don't use pork, which is too

▲ Herb-seasoned tomatoes in an electric dehydrator.

▼ Dried apples.

fatty. Cut into trimmed strips around an inch wide and half an inch tall, and as along as you want; cut along the grain. Sprinkle with ground pepper and salt.

2. Normally, when you dehydrate foods, you would want to soak fruits and vegetables in a solution of vitamin C or sugar for five minutes to prevent oxidation and discoloration. However, this extra soaking will make it difficult to dry foods in the sun. If you are using an electric dehydrator, get a big bowl or bucket of ice water and dump some sugar or citric acid (vitamin C) in it. As you chop, put the finished slices into the bowl until you are ready to lay them on the drying trays.

3. Spread one layer on a drying tray. When making fruit leather, line the tray with plastic wrap or even parchment paper. Each type of food will dry at a different rate, so it's important to keep them all separate. Make sure you label everything so you know what it is.

4. Put the trays in the hot sun or in the electric dryer. If you are drying meat in the sun rather than the dryer, it should be on a tray that is 4 feet above a slow fire. The firewood should be non-resinous hardwood and with very low flames because its purpose is to keep away the birds and the flies. Green wood that makes a lot of smoke works well for this.

5. Turn big chunks of food three times a day and small foods once or twice a day. In a dehydrator, move the trays of almost-dry food to the top and the moist food to the bottom. It should not take more than two days for them to dry. Everything needs to be protected from dew and bugs. If you must continue the next day, bring everything inside before dusk and bring the food out again in the morning when the sun is out.

6. Vegetables are dry when they are brittle and break when bent. Fruits are leathery or brittle and should produce no moisture when squeezed. Fruit leathers will be a little sticky but easily peeled from the paper or plastic lining in the tray. Pods will become dry and brittle. Meat should be hard and solid, with a uniform appearance; this means that they should be the same dark, red color throughout, without any excessively large wrinkles in one place. Break a piece to make sure it is dried in the center.

7. Put fruits and vegetables into a wide-mouthed bowl for a week, stirring it two to three times a day. Keep it covered with a screen or porous cloth. This conditions the food to resist mold. Then repack it tightly in an airtight container or freezer bag and store in a dark, dry place.

8. If you want to pasteurize the food, put it in the oven for 30 minutes at 175°F (79°C). Remember to label with the type and date, and check it in the first two weeks for moisture—if you find that there is some, you'll need to dry them some more. Properly dried food should stay good for at least six months or more. If you find bugs, remove the bugs and roast the food at 300°F (149°C) for 30 minutes.

NO-ENERGY STORAGE

**Nothing would give up life:
Even the dirt kept breathing a
small breath.**

~ Theodore Roethke

Live Storage

Pumpkins, potatoes, dry beans and
peas, onions, parsnips, turnips, apples,
oranges, pears, tomatoes, and most
other root vegetables can be stored
live, or without any processing, in a
properly maintained cold storage. While
the conditions of your cold storage are
essential to your success in this, picking
a species made for preserving this way
and harvesting at the right time are also
extremely important. Make sure that
the foods you use are not bruised or
blemished in any way and remove the tops.

The first step is to leave the dirt on,
which will help protect them from decay.

Use plastic buckets or enamel cans, as these
will be rodent and decay-proof. Fruits need
to be stored away from vegetables because
the gas produced by apples can cause
vegetables to sprout. Pack root vegetables
in damp sawdust, sand, or moss. Keep
potatoes out of any light, or they will turn
green and become poisonous to humans.

Put a thermometer on the inside
and outside of the cellar and monitor
the temperature every day. Use doors
and windows to maintain a temperature
of 32°F: Open the door in cold weather
and close the door in very cold or hot
weather. Alternatively, you could install a
fan attached to a thermostat that functions
similarly to the kind used to ventilate a
greenhouse. The food needs humidity so
that it doesn't dry out, usually 60–75%. If
necessary, you can set out pans of water,
sprinkle the floor with water, or cover the
floor with damp sawdust. If it is too damp,
take pumpkins, squash, and onions to a
drier area or they will rot quickly. Remove
all spoiled food, and if something is about

Food	How long	Method
Apples (especially Winesape, Granny Smith, Black Arkansas, Idared, Liberty)	4 months	In small crates stacked no more than 2–3 high. Stack the ripest ones on top. Store above ground on a shelf or table.
Cabbage	3 months	Pick before frost, removing roots and outer leaves. Place upside down in a single, loose layer in a crate. Stack the boxes and cover with a tarp.
Carrots	4 months	Line crate with leaves, stack carrots upright against each other. Stack crates above ground.
Chestnuts	6 months	Soak nuts in water for 2 days. Remove anything that floats. Let dry for 1 day on a screen out of the sun. Store in bucket with sand with a screen cover.
Leeks	Varies	Cut off roots and leaves. Transplant to a container of sand or sawdust and water once during winter.
Root vegetables	Varies	Root vegetables need to be put into any large waterproof container and layered with sand or sawdust so they are not touching each other.
Squash	3- 8 months	Wipe down with vegetable oil and wrap loosely in newspaper. Throw out moldy ones.
Tomatoes	4 months	Pull up the entire plant at the start of autumn. Wrap each tomato in newspaper and hang the plant upside down. The green ones will ripen.

to spoil, dry it quickly before it rots. If something is rotting or molding, get rid of it immediately, and make sure there are no insects infesting anything.

If you can't build a cellar or cold storage, you can use a *clamp*. This is a very old device that is something between using a cellar and simply leaving roots in the ground. It can range from just a hole in the ground, to an old washing machine drum, to a nice brick box sunk into the ground. The pit can be between 8–20 inches (20–50 cm) deep and lined with something to stop rodents. This can be wire mesh, clay, or brick. Inside, throw down a layer of sand, leaves, straw, or twigs and begin layering the vegetables with layers of dry material in-between. If you don't have very many to store, just put them all in one pit (not packed too tightly), but if you have lots, then make more than one clamp, one for each type of food. In the center leave a hole or tunnel up to the top and fill it with twigs for ventilation. Cover the top with another layer of dry material and cover that with a wooden board. Cover that with plastic and put a heavy rock over the top to keep animals out.

Leaving Them in the Ground

Some root vegetables can simply be left in the ground. This is similar to using a clamp except that they are just individually surrounded by the soil they grew in. They must be protected from frost, however, and this is done in different ways depending on the vegetable. You can do this in October or November, before the first frost. If you have raised wooden beds, then you simply need to cover the plants in the manner described below, but if you do not, you will need to sink wooden boards into the ground around the bed to help protect the plants.

FERMENTATION

I put everything I can into the mulberry of my mind and hope that it is going to ferment and make a decent wine. How that process happens, I'm sorry to tell you I can't describe.

~ John Hurt

The Safety of Fermentation

Lacto-fermentation has become more popular again in recent years because it saves the nutritional properties of the food preserved and has all kinds of friendly bacteria. Where other types of food preservation techniques try to kill all the bacteria, fermentation encourages it. It works because the fermentation process produces lactic acid, which kills botulism

Beet root	Leave in the garden until very cold.
Brussels sprouts	Cover well with dry straw and a sheet of plastic.
Cabbage	Dig an 8 x 8 inch trench running east to west. Lay down a row of cabbages in the trench with the stem towards the south. Cover with straw.
Carrot	Cover well with dry straw and a sheet of plastic.
Cauliflower	Leave in the garden until very cold.
Chicory	Cover well with dry straw and a sheet of plastic. Prevent rot by uncovering in mild weather.
Curly kale	Cover well with dry straw and a sheet of plastic.
Endive	Cut off the leaves, cover with 8 inches of dirt. Cover shoots with more dirt. Eat in early spring.
Jerusalem artichoke	Leave in the ground and cover with straw.
Kohlrabi	Leave in the garden.
Leek	Cover well with dry straw and a sheet of plastic.
Lettuce	Dig a 16 x 16 inch trench and lay the heads in not touching. Cover with straw.
Parsnip	Leave in the ground and cover with straw.
Radish	Cover well with dry straw and a sheet of plastic.
Salsify	Leave in the ground and cover with straw.
Turnip	Leave in the ground and cover with straw.

and other bacteria. For this reason, it can be much safer than canning or even eating raw vegetables, which can harbor E. coli.

It is highly recommended that you put the fermenting foods into jars with rubber-sealed lids. The rubber seals release gasses that build up during the fermentation process, preventing an explosion. Traditionally, people used crocks as well. You can sterilize the jars by pouring boiling water in them if it makes you feel better, but soap and water are enough. Grow the food yourself or get it from a farmer who has clean produce and a good reputation.

Kimchi

One of the most popular foods in the world is fermented cabbage, also known as *kimchi*. Kimchi recipes vary, but they all have several ingredients in common:

1 Napa cabbage cut up into 2-inch cubes or wedges

¼ cup sea salt

garlic, ½ bulb per cabbage

1 shredded radish

1 onion

lots of chili powder or a little cayenne powder

½ teaspoon of sugar

2 tablespoons of unchlorinated water

ginger

The cabbage cubes are thrown into a bowl and completely covered and mixed with the salt. This should be left to stand for at least two hours. Then you can wash off the cabbage if you want. The garlic, ginger, onion, and water are mashed up or blended together, or the powders

 Kimchi.

mixed with a little sesame oil, and stirred with the sauce, radish, chili, or cayenne. Pack tightly into extremely clean jars with rubber-sealed lids and keep in a dark cool pantry for a couple of days. During that time, check on it periodically and smash it down inside the jar. Smashing is important, as you need to keep the cabbage at the top well covered by the liquid. It will begin bubbling, at which point you can put it in the fridge or cold storage. It can be eaten raw for three weeks, and after that you will have to use it cooked with something because it will be too strong.

Sauerkraut

Sauerkraut is extremely similar to kimchi because it was inspired by kimchi. The recipes also vary greatly, but at its most basic level you just follow the same steps as for kimchi.

1 cabbage, or several types of cabbage to equal the same amount

▾ **Sauerkraut.**

¼ cup sea salt

2 tablespoons unchlorinated water

Chop up the cabbage. You can also add chopped turnips, beets, greens, Brussels sprouts, apple, herbs, and spices. Pack it tightly into your (very clean) jars, adding the salt as you go. When you smash the cabbage, it forces some of the liquid out. Top off each jar with a spoonful of salt and add a few spoonfuls of hot water. Seal the lids tightly with rubber seal lids and let the jars stand in the kitchen for a few days. The cabbage will ferment, and the water level should rise, at which point you need to smash the cabbage down as hard as you can a couple of times a day. If you see mold, scrape it off. After four or five days, put it in the cold storage or cupboard and let it sit for four weeks. Then eat.

Pickling

Cucumbers are the most popular type of pickle, but beets, carrots, green beans, onions, radishes, Swiss chard ribs, turnips, zucchini, and many other vegetables can be pickled too. The recipe is pretty much the same as for kimchi and sauerkraut, with only a few variations. The recipe is as follows:

1 pound of sliced cucumbers
1 cup of sea salt for salting
peppercorns
1 ½ tablespoons sea salt for brine
1 cup of unchlorinated water
mustard seeds
lots of fresh dill
2 cloves chopped or mashed garlic

Lay the cucumbers tightly into a bowl, adding salt to each layer. Fill with water so that there is at least an inch of water on the top. Soak the dill heads upside down in salt water as well. After 24 hours mix the water and brine salt so the salt is completely

The Ultimate Guide to Permaculture

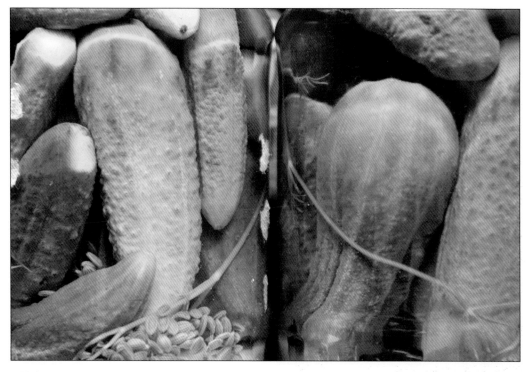

▴ **Pickles.**

dissolved. If the cucumbers still have peels on them, you can poke holes in the peels with a fork. Pack the cucumbers tightly into extremely clean or sterilized jars, layering them with the mustard seeds, peppercorns, dill, and garlic. Don't fill the jars right to the brim. Optionally, you can place a horseradish leaf on the top to protect the top layer. Close the jar tightly and keep it in the kitchen for a couple of days so you can watch it. When it begins to form bubbles on the top, put it in the fridge or cellar for 6 weeks before eating.

Yogurt

Yogurt isn't fermented in the same way that sauerkraut or pickles are made, but it is technically fermented milk. Bacteria turn the lactose into lactic acid. While store-bought yogurt has flavoring and sugar in it, homemade yogurt is simply milk and starter. You can use any milk, and the starter can simply be plain yogurt that has *active cultures* in it. If you get the yogurt from the store, it will use that phrase on the label.

To make the yogurt, use a thermometer to precisely heat a quart of milk and 3 tablespoons of plain yogurt to 100°F (38°C), then turn off the heat and keep it warm for 8–12 hours. You can use a wood stove, thermos, an oven, the sun, or a crockpot. With a thermos and the sun, you can heat the milk directly before letting it sit. With other methods, preheat the oven or crockpot beforehand and then shut it off when the milk is sitting. You may have to turn it on periodically just to keep the temperature warm. On a woodstove simply set it at the back when the stove is cooling. The yogurt is done when you can tilt it and it doesn't run.

Plain yogurt is a tough transition for many people. It can be used in sauces or as a replacement for sour cream, but if

you want to have sweetened yogurt, you can use sugar, honey, or fruit syrup to make it more palatable. If you do flavor it, remember to always keep a little bit of plain yogurt set aside in the fridge.

Sourdough

Sourdough is a way of fermenting your own yeast for bread. The yeasts are wild and caught from the air. All it takes is a little flour, some water, and time. Mix half a cup of any type of wheat flour, and 1–2 tablespoons of unchlorinated water. Knead this into a small piece of dough. Put the dough in a glass or ceramic (not metal) jar or bowl covered with a damp cloth or cheesecloth for two days. In that time will form a hard crust, which will be slightly wrinkled.

Day 3: Pull the crust off, add one cup of flour and a little water and cover the jar again for 24 hours.

▲ Homemade yogurt.

▼ Sourdough bread.

The Ultimate Guide to Permaculture

Day 4: Refresh the dough again by removing the hard crust, adding another cup of flour and some more water to form another blob of dough.

Day 5: After 8–12 hours the dough will have risen, and if you poke it, the hole won't spring back.

This process of feeding the starter can be difficult to remember. If you simply check it every morning for four days and add a cup of flour and about half a cup of water each time, you should have sourdough starter by the fifth day.

You can use the starter in any sourdough bread recipe. Put the starter in the fridge until you want to use it, but never use the whole thing. There are sourdough starters in the world that have been used and passed on for over 100 years. To make more starter, simply repeat the process of feeding and refreshing your existing starter on the kitchen counter again.

To make bread, you will need:
a handful of starter + 2 cups flour + 2 cups water

2 cups flour
1 tablespoon of salt
1 cup of warm water
a little oil

It is best to start this bread in the morning the day before you need it. Mix the starter + flour + water and set the mixture on the counter covered with a damp towel all day. Since you had it in the fridge, this will get it going again. In the evening, mix the rest of the flour, salt, and water together into a nice dough. Knead it very well and let it rise overnight in the biggest bowl you have, covered with a towel. In the morning, punch it down and divide it in half. Form into two round loaves and let them rise for 3–6 hours. Preheat the oven to 450°F (232°C) and put the loaves on a baking pan. Slash the tops so the bread can expand and spray them with a little water. Spray the inside of the oven with more water. Bake the loaves for about 30 minutes or until they are nicely browned and sound hollow when you tap on them. Allow to cool before you cut them.

Alternatively, to use less energy, you can make sourdough in a crockpot by simply putting the dough on a small wire rack into the pot and adding some water to the bottom. Close the lid, and the steam from the water will bake the bread. Set on low for four hours. This method can also be used in a solar or thermal cooker as well.

SUGAR

I pity them greatly, but I must be mum, for how could we do without sugar and rum?

~ *William Cowper*

Jam Without Added Pectin

Pectin can be bought from the store, but it is much better to use the natural pectin found in apples. To do this, save the peels and cores of apples and tie them up in a bundle of unbleached muslin cloth. The more apples you put in there, the more pectin you will have, but even a few will make a difference. There is also a distinction between jams and jellies. Jelly is the kind of spread that has no chunks of fruit in it. Instead, the fruit has been pureed and strained. Since you are using natural pectin, you have to make jam. The fruit in jam has simply been crushed or just boiled until it has turned into mush. The measure of sugar has everything to do with the quality of your jam. Some fruits also need extra acidity to form a gel. Usually this is lemon juice, but any citrus could work.

Does it have to be sugar?

Sugar as a preserving method is almost exclusively used for fruit because of all the natural sugars they already have. The extra cane sugar has the function of preserving the color of the fruit and also helps to create the firm consistency we are familiar with in jams and jellies. It also helps the fruit last longer; without the sugar, the jam won't store as long. It doesn't have to be cane sugar, but cane sugar ensures success. A note about honey: It is recommended to substitute only up to half the sugar, or your jam will be a runny syrup that lasts a much shorter amount of time.

To substitute other sugars, use:
- Honey: ¾ cup for every cup of sugar
- Brown sugar: use the same amount as sugar, but it has a strong flavor—works well for peaches
- Raw sugar: add ¼ cup more for every cup of sugar

Fruit	Sugar	Citrus Juice
1 cup apples	¼ cup	1 ½ teaspoons
1 cup apricots	1 cup	2 tablespoons
1 cup berries	¾ cup	1 ½ teaspoons (optional)
1 cup grapes	1 cup	
1 cup peaches	¾ cup	2 tablespoons
1 cup pears	¾ cup	2 tablespoons
1 cup plums	¾ cup	
1 cup rosehips	¾ cup	

Clean, peel, and remove the stems of the fruit you want to add. Grapes need a little extra processing to remove the peel, which can be done by simmering or squeezing. Rosehips are gathered in the winter, after the first frost, and must be pureed to remove all of the seeds. Once they are all prepared, put all of the ingredients in a big pot with the muslin bag of apple cores, with the exception of the rosehips, which don't need apple cores to thicken up. If you are making apple jam (or apple butter), you will have to add 1 tablespoon of water per 1 cup of apples. Bring the fruit to a simmer over low to medium heat, stirring now and then. Towards the end you will have to stir more. Continue until it has reached a consistency that doesn't drip. Most jams, especially berry jams, may still be fairly liquid and won't have that store-bought jiggle, but as long as the jam slides slowly off a spoon, it should be done.

Fill the jars up as much as you can with the hot jam, close tightly, and turn the jars upside down. The jam will help sterilize the empty space at the top. Store them upside down, and they will last the winter. Other ingredients to add include cinnamon, fresh walnuts, raisins, mint, currants, hazelnuts, and vanilla.

Chutney

Chutney originated in India and is usually prepared right before a meal, but it can also be used to preserve any kind of fruit or vegetable. It is very similar to jam, but there are a bunch of different spices and flavors combined together, and rather than being something reserved for toast, chutney is used to top cold meat, potatoes, rice, grain, and salads.

▾ **Chutney.**

You will need:

4 cups of chopped fruit or vegetable (apples, mangoes, plums, tomatoes, rhubarb, etc.)

1 cup of a complimentary vegetable (radish, zucchini, eggplant, etc.)

1 cup of chopped onions

2 tablespoons of brown sugar

½ cup vinegar

salt

Herbs (ginger, mustard seed, cloves, cayenne pepper, rosemary, pepper, curry, etc.)

Throw all of the fruits and vegetables and onions into a pot with the herbs and salt to taste, add a little water, and bring the mixture to a boil. Simmer until everything is very soft and mixed together. Add the sugar and vinegar and continue boiling until it has the same consistency as jam. Sterilize the jars and lids and pour the chutney in while it is very hot. Close the lids immediately and store upside down. Ketchup is basically a chutney.

7 | The Zones

For priority in location, we need to first attend to Zone 1 and Zone 2; these support the household and save the most expense. What is perhaps of greatest importance, and cannot be too highly stressed, is the need to develop very compact systems. . . . We can all make a very good four meters square garden, where we may fail to do so in 40 square meters.

~ Bill Mollison

ZONE 1: DESIGN

Prowling his own quiet backyard or asleep by the fire, he is still only a whisker away from the wilds.

~ Jean Burden

Elements found in Zone 1:

- Greenhouse and shaderoom
- Garden shed
- Workshop
- Herb spirals
- Kitchen garden with salad for clipping and pathside vegetables
- Miniature fruit trees
- Broad beds with plants like leeks
- Broadcast sown grain
- Vines and trellis
- Semi-wild and hardy trees and bushes
- Very quiet small animals like rabbits and quail
- Compost bin
- Clothesline
- Outdoor kitchen or barbecue pit
- Pigeon, rabbit, or quail pens

Design Considerations of Zone 1

Zone 1 hugs the house because it is the area you will work in or walk through every day. The size and shape of it depends on how much land you have, how you access it, your own daily schedule, how much time you have, and whether you have animals in a barn located at the edge in Zone 2. City dwellers may have a Zone 1 only reaching 10 feet, while rural people might have a half-acre or more. The zone is designed with the same considerations that you used in chapter 1 to plan the layout of your property, such as the climate, wind, and relationships between the elements.

On top of those considerations, you also need to keep in mind:
- How will I access each element? Do I need a road to get to Zone 2 or beyond?
- Where is the best place for the road? When I build the drainage from the paths and roads, where will the water runoff go?

- Where are the doors to the house, and where will the greenhouse and shaderoom go?
- Where is the best place to put a clothesline so that it gets sunlight and also has easy access to the laundry room?
- Where is the best place for a play area? Even if you don't have children, a place to relax or play an outdoor game is important.
- If you are using wood heat, where will the woodpile go so that it is close to the door and sheltered from the weather? Where will the chopping block be?
- What kind of outdoor kitchen or barbecue are you building? It will probably extend from the shaderoom.
- Where is the water source for this zone? Where are the storage tanks and hoses, and where will you distribute the graywater from the house? What kind of irrigation will you use?
- What kind of fencing or hedge will span the perimeter to keep out larger animals?
- Are you growing pigeons, bees, quail, rabbits, or worms? What kind of housing will they need? How will you give them water? Where will the waste go?

The Ideal Zone 1

When your garden is in its best working order, the soil will be completely mulched and the soil rich with organic matter. When you harvest a plant, as much of it will be eaten as possible, and the rest composted in the compost bin. Some of your dill, fennel, and carrots are left to go to seed to attract parasitic wasps. In the winter, a green manure crop is grown and turned over to return precious nutrients to the soil. Any volunteer tomatoes and

▾ Zone 1 is highly controlled, irrigated, and drained.

The Ultimate Guide to Permaculture

▲ Zone 1 irrigation and controlled garden beds.

▼ Zone 1: garden beds, Zone 2: fruit trellising, Zone 3: forest garden.

cucumbers from the compost heap are replanted in an empty spot.

The gardens are self-sufficient, and the trees provide shade and food just a few steps from the kitchen. A small grassy area extends from a shady trellis that covers an outdoor kitchen with a mud oven and a sturdy table where people can gather for delicious food and conversation at dusk in the cool air of the evening. A pathway leads from the backdoor through winding paths sheltered on either side by an abundance of beautiful and edible plants that are picked when needed. The path leads to a

▲ The ideal outdoor kitchen and living space.

small pond where a few ducks splash in the water, eating insects and helping the fish to grow. They hide their nest among the reeds, which you gather later and make into baskets.

The first concern with the plants directly next to the house is how they can make the house cooler in the summer and warmer in the winter. *Deciduous* (trees that lose their leaves in the fall) varieties are placed on the sunny and eastern sides of

▾ Deciduous trees provide shade in summer, allow heat in winter.

the house. In the winter the sun will be able to shine in, and in the summer the leaves will block the sun. The shade room and all sides of the house can be grown with vines (like grapes) while the trees are growing. Around the other sides of the house (besides the east side) grow evergreen trees to protect the house from the heat and wind at all times of the year. Bamboo is a particularly valuable windbreak, especially in a hurricane zone.

Herbs should be located as close to your kitchen door as possible, so you can run out and grab what you need as you are cooking. This is accomplished with the herb spiral. At the top of the spiral are the thyme, rosemary, and sage, and the shady places are for mint, cilantro, parsley, and chives. At the very bottom is a tiny pond lined with plastic for water chestnut or watercress. A sprinkler at the top waters the whole thing.

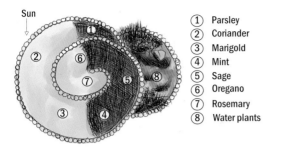

①	Parsley
②	Coriander
③	Marigold
④	Mint
⑤	Sage
⑥	Oregano
⑦	Rosemary
⑧	Water plants

▲ Spiral herb garden with small pond.

Located near the herb spiral is the salad bed. This is a narrow bed for easy reach from the path, where more herbs, salad greens, chives, and shallots are grown. These types of greens grow very quickly as they are trimmed, and the soil should be kept mulched throughout the year.

Any pathways around the house are populated on each side with vegetables that can be harvested throughout the summer. These are transplanted from the greenhouse and include Swiss chard, Brussels sprouts, onions, celery, broccoli, kale, mustard, spinach, peppers, zucchini, and fennel. Most of these can be picked as needed, but if you do use it up, you can have seedlings growing in the meantime to be transplanted when ready. At the end of the season leave some to go to seed for next year.

The main garden is made up of two types of beds: narrow and wide. Beans, tomatoes, carrots, peas, zucchini, eggplant, and herbs such as chamomile and cumin can be grown in the narrow beds. These plants are picked more frequently and need easy access. The wide beds are for vegetables and fruit that mature all summer and are harvested once, such as corn, melons, onions, potatoes, beets, leeks, and turnips. They can be planted close together and mulch themselves. They usually take less work to maintain, and so you may want to have additional wide beds in Zone 2 as a cash crop.

▼ Beets.

At the edge of Zone 1, and possibly around some of the areas of the gardens or outbuildings, hedges are grown to keep out animals, the wind, and weeds. These can also perform other valuable functions like creating mulch material, providing animal food, fixing nitrogen, and creating food for humans. Plant your hedges and mulch the soil around them heavily with cardboard, straw, or sawdust. While you wait for them to grow, you will also have to grow a quick growing plant that can stop the grass from invading the area again. Comfrey, bamboo, Jerusalem artichoke, or Siberian pea shrub can all provide this type of temporary barrier. Plant them in a band 4 feet wide outside the hedge bed and keep them controlled so that they don't spread, which some of them tend to do. If you want just a small barrier within a garden, you can use most perennial herbs, such as rosemary.

If you have strong winds and need to protect the garden before your hedges grow, you can build a barrier instantly with tires, which also act as a thermal mass that absorbs heat. The ground should be prepared first with newspaper and mulch. Then the tires can be stacked and filled with earth, compost, hay, or whatever scraps you may have. At the top you can plant a species that is wind tolerant.

RABBITS

What is a country without rabbits and partridges? They are among the most simple and indigenous animal products; ancient and venerable families known to antiquity as to modern times; of the very hue and substance of Nature, nearest allied to leaves and to the ground.

~ Henry David Thoreau

How They Fit into the System

In a permaculture system, bunnies are not kept as pets. They provide meat, manure, and fur. Unlike other animals, they can never be allowed to roam the gardens, because they will eat everything and multiply beyond control. Instead, they live in a hutch with a mesh floor so that their droppings will fall down for easy cleanup. A worm bin can be placed underneath for even easier cleanup. Rabbits eat grass, leaves, twigs, hay, vegetables and kitchen scraps, and they are especially practical for urban homesteaders because rabbit hutches are legal in most places. Any rabbit breed can be eaten, but there are rabbits that are designed specifically for meat and grow to be much larger, such as California, New Zealand, Champagne d'Argent, or Florida White.

The nutritional value of rabbit is not as high as that of chicken or other meat staples, but since rabbits are very self-reliant and quiet, they make a valuable meat source. They need a constant supply of fresh water, and the wire mesh of their cage should not be larger than half an inch. The hutch should be kept clean, dry, and sheltered from the weather.

Breeding

Rabbits breed very prolifically and easily but may not always be the best mothers. Small breeds should not have babies until they are six months old, and a large breed not until they are nine months old. Males and females should be kept separate and only put together for breeding for a short time under supervision by putting the doe into the male's cage so that he notices her. It's a good idea to have him try again about eight hours later just to make sure. Once she is pregnant, she should be provided

▲ **Baby rabbits or kits.**

with a nesting box and soft material such as hay or down. After the babies are weaned, the males and females should be separated so that they will not breed too early. They become very territorial and will need their own cages shortly anyway. It should be noted here that while siblings should not mate, it is common practice among rabbit breeders to allow mother and son or cousins to breed. Don't breed a rabbit with defects or illness, and choose a good breeding pair promoting the best traits. Around 30 days later, the doe will *kindle* or give birth.

A doe needs a nesting box in her cage at least a few days before she kindles, which is a low-sided wood box with a tall back and a roof awning that comes down over the back. She will pull her own fur out to make a nest. If she doesn't, you can gently pull small bits of fur from all over her body. A couple of days before birth, her fur will

loosen, and this will be easy to do. You can also add clean straw for even more warmth. Sometimes does don't give birth in the box, and you will have to move the babies back into the nest. They will probably be too cold to survive, but you can try to warm them up. If you are able to save them, put them back in the nest in a little hollow of fur and cover them. Sometimes a doe might eat her young, which can be caused by anything from stress to poor nutrition to disturbing her when she was about to give birth. She may just be a terrible mother. Sometimes does don't feed their young, or they step on them and kill the babies accidently. It is possible to try to feed, them but chances are they won't survive. Since she can have another litter soon, it may not be worth the trouble of feeding the babies every two hours. It is typical to breed a doe every six weeks and wean the babies when they are around five weeks old, and with such an intense breeding schedule you may be able to raise 300 pounds of meat.

PIGEONS AND QUAIL

And it came to pass, that at even the quails came up, and covered the camp: and in the morning the dew lay round about the host.

~ Exodus 16:13

How They Fit into the System

Pigeons are kept in cages tall enough for you to walk around inside. Quail can be kept in much smaller cages and up to six can be raised in a square foot (0.09 meters), although for our purposes we would want to give them more space than that. Quail can also live in the greenhouse because they don't eat the plants; pigeons eat seeds and grain, and quail eat insects. They provide eggs and meat and like rabbits can be legally grown in the city. For people who live in urban locations and are not able to raise chickens, pigeons, and quail are sometimes allowed (although not necessarily in the quantities you will want to raise). Quail are considered wildlife, and in many places you may need to get a game bird license. A breeding pair of pigeons can produce twelve *squabs*, or baby pigeons, per year. Squabs are considered a gourmet dish and are incredibly easy to raise. Quail lay about 200 eggs a year (almost every day), depending on how much light they have. If you add lighting during the winter, they can produce 300 or more. Unlike pigeons, quail aren't very good at brooding their own eggs, and like chickens, they need a little extra help. They are more often raised for the eggs than for

their meat, because they are smaller than pigeons but lay more eggs.

The pigeon coop should be at least 6 × 8 feet (1.8 x 2.5 meters), and 7 feet (2 meters) tall. Each breeding pair also needs a nesting box attached to the wall off the ground, filled with straw or hay, and a constant supply of fresh water. Pigeons and quail are social animals, and so it is better to have at least three breeding pairs at any one time. There are pigeon breeds just for show, but you are looking for a large meat variety like Cropper, White King, or Silver King. Quail breeds raised for meat include Coturnix (Japanese) and Eastern Bobwhite.

Pigeons and quail can both be allowed to fly free during the day because they will return to the coop at night. They will eat from the garden and fertilize everything, and it is a common practice. There is a much higher chance that predators will eat them if you do this, but make sure that they get closed in securely at night when they return to the nest. While most coops are designed to walk into, you could also design a coop that sits off the ground like a rabbit hutch to deter possums or weasels.

Breeding

Pigeons mate for life and need very little care. As long as you have a breeding pair together, they will find a nesting box and settle in to make babies. Both take turns sitting on the eggs, which will hatch in

▲ **Quail house.**

▲ Nesting pigeon.

18 days. Coturnix quail eggs will hatch in 18 days, and Bobwhite in 23 days. Pigeons will feed the squabs regurgitated food, and after 28 days they can be butchered, before they start to fly. Quail can be butchered in 6 weeks. To kill a squab or quail, take it from the nest in the morning before it eats, cut off the head, and hang it to bleed out (the same as you would a chicken). The feathers are carefully pulled out without scalding. Remove the feet and throw them out, then cut the body from the vent to the breastbone and remove the organs. Save the gizzard, heart, and liver. Rinse with cold water and refrigerate or freeze as soon as possible. It takes two squabs to make a meal for one person.

It is difficult to tell which birds are male and which are female. You will want to eventually replace your first breeding pairs with younger pairs, and so you will have to watch their behavior. Males are noisy and busy strutting, while females are very quiet and sit still.

BEES

For so work the honey-bees, creatures that by a rule in nature teach the act of order to a peopled kingdom.

~ *William Shakespeare*

How They Fit into the System

Bees are the producers of most of what you eat. Without their pollination, producing enough food to feed us would be impossible. They also make honey and

As a beekeeper you are certain to get stung many times, and you can build up immunity. However, you can also suddenly have an allergic reaction. Before buying bees, get tested for allergies to bees and bee stings. Buy protective bee gear and always work with someone so that if you do develop an allergy, your partner can get help. When you get stung, scrape the stinger out quickly with your fingernail so that less venom will enter your skin.

beeswax. The most difficult part of keeping bees is making sure they have enough forage to make enough food to keep them alive through the winter. If they don't have enough, they have to be moved or sugar water added to the hive to try to keep them going. To choose a location for the hives, we have to go back and think about our sectors. Bees prefer to fly at least 300 feet to their food source, and they won't forage well in the face of a cold wind. Using the sector map, place the hives away from the wind and use hedges of herbs to shelter them in the direction that you want them to go.

There are two kinds of forage, pollen and nectar, and bees need both. The pollen

▾ Bees fly outside the circle for food.

species are planted within 100 feet of the hives, and the nectar species are planted at least 300 feet or more away. The line of herb hedges doesn't even have to be more than 3 feet tall, and it directs them from the hive doorway towards the forage by sheltering them from the wind. These can be rosemary, acacia, or built up soil beds planted with thyme, catmint, or field daisies. The pollen producers around the house you have probably already planted to provide shade can be willow, acacia, pine, and vines like grapes. Everything else can be planned to flower in succession so that the bees can have a constant supply throughout the season. Having a minimum of 30 species to forage from is insurance for your hives. These include gooseberries, apples, white clover, blackberries, citrus, buckwheat, mustard, and other fragrant herbs. The rest can be supplied from field crops.

There are three types of honeybees: Italian, Caucasian, and Carniolan. Italians work harder, Caucasians sting less, and Carniolans are the gentlest. Honeybees can only sting once (unlike wasps) because they die, and this makes them less likely to sting. You can either buy bees from a supplier or buy a whole hive from a local beekeeper. The last option is the easiest because the hive will be well established. Once you have one or two hives, you can have an unlimited supply by encouraging bees to establish new hives.

The Tools

You will need a hive, a smoker, a *hive tool* (a small hooked lever for taking frames out of the hive), bee clothing, a bee brush (for brushing bees off a frame), and a feeder. A hive has several layers. At the bottom is a *hive stand*, a platform that makes sure the hive is level. Above the stand is the

bottom board, a thin frame that holds up the *brood chamber.* The brood chamber is where the bees make their home, and it is where the queen lays eggs, which are deposited in cells to become baby bees. *Supers,* or honey supers, are shallow boxes that sit on the brood chamber and usually hold honey, although sometimes they have baby bee cells. It is better to have a shallow super than a deep one because the latter can get too full of honey and difficult to carry. Each super has 10 vertical removable frames. On each frame is a *foundation,* a flat sheet of beeswax that has hexagons imprinted on it as a template on which the bees to build cells. In most areas used bee equipment is illegal because of disease, so contact your area's department of agriculture before purchasing any.

▲ Hive with two supers.

▼ A hive tool.

Handling bees:

1. In your smoker, start a fire with crumbled paper and add tinder such as pine needles and dry grass. The fuel doesn't need to be too dry, because you want it to create smoke. When the fire is burning well, close the lid and use the pump to keep it smoldering.

2. Stand to one side of the entrance to the hive and blow smoke in the door. Wait a minute or two, take off the cover, and blow more smoke in the top.

3. Anytime that the bees start to get agitated with you, use more smoke. Be careful not to hurt a bee, or it will release a panic odor alerting the bees to sting you.

   The hive can be inspected if the temperature is over 50°F (10°C) and the weather is nice. Some people take a look once a week, but that's a bit more than

Knowing the hive

Brood cells: have dark colored caps (unlike honey cells which are light colored) and contain baby bees.

Queen cells: are 1 inch (2.5 cm) long and look like a peanut shell that hangs away from the rest of the comb. They contain baby queens.

Drone cells: stick out like the queen cell, but not as far, and have bullet-shaped tops. They contain baby drones.

Worker cells: are the smallest cells, are level with the rest of the comb, and contain baby workers.

Queen bee: the queen is unique. She is 1 inch (2.5 cm) long and has a tapered body. The other bees won't crowd around her.

Drone bee: they don't have stingers, are very fat, and have big eyes. Their only job is to compete to mate with the queen. It takes 24 days for a drone to hatch.

Worker bees: they are the ones that sting, and they keep the hive going. It takes 21 days for them to hatch.

▲ **Bee smoker.**

the bees are comfortable with. In general it is only necessary to check in if you suspect a problem, and in the spring and fall. In the spring carefully remove every single frame and find the queen. Look for queen cells, find out how many bees there are, how many brood cells there are and what type, how much honey is coming, and whether the bees need more supers. Supers prevent overcrowding, which prevents swarming.

Bees always need fresh, clean water. Big ponds don't make a good water source, however, because bees may drown or be killed by a resident dragonfly. One or two hives only require an outside faucet left to drip onto a slanted board. If you have many hives, then either a soaked mat near the pond or a very tiny pond near the hive can supply them. If you see bees standing around outside the door of the hive in warm weather and you know the hive has a high population, it means they are having

trouble cooling the hive. Move the hive into the shade, make the entrance larger, and stagger the supers for ventilation. In the winter, making the door much smaller will help keep heat in and also prevent mice from creeping in and stealing honey.

Bees fare better in a warmer climate because of the greater availability of food and because they are less likely to freeze. If you get a gallon per summer per hive in the first three years, you will be very lucky because it takes a while for the hive to get established. With practice you should eventually get 4-5 gallons. Placement is the key to a successful hive. Point the door of the hive towards wherever you want them to go and away from houses, barns, and loud motors. One hive will need 50–100 pounds of honey to get through the winter. If the bees get low on honey, feed them 2 parts granulated sugar per 1 part water, and in the spring you can give them artificial pollen.

Making New Hives

To move an entire hive, plug the door of the hive very tightly with a porous material that allows air in, so the bees won't suffocate. It must be very secure or you will find yourself in the middle of an angry swarm. To create more hives, at the beginning of May you can take four frames with brood cells from your most established hive, as well as some honey and some *bee bread* (or pollen). Bee bread is yellow and grainy. You will also need worker bees, which you can just brush into the hive. It is better to have one queen cell per frame, but if you don't, they will make one. Put the frames into the new hive and stuff the door loosely with grass. It can take two weeks for the hive to produce a queen, and then the queen needs four weeks to mature and mate. The bees will put so much effort into this process that they will only make enough honey to support themselves over the winter, and you will not be able to collect any honey that year from the source hive or the new hive for yourself. You also run the risk of losing a new colony if the bees fail to feed royal jelly to the queen on the first day of hatching.

Beekeeping calendar:

Early spring: Check that they have enough food and supply them with artificial

pollen. In some cold climates you might see dead bees at the bottom of the hive, but this usually means the queen has died. If the queen dies and there are no eggs, the worker bees will wander around and eventually die. If they do have eggs or larvae, they will make a new queen. The workers should keep the hive very clean. If it gets dirty, it's a sign that the queen is gone and they will all die.

Late spring/early summer: When you think the bee population is big enough, add another section to the hive to hold the increase in comb production. This is also the time to split the hive into two hives if you want. If the bees feel too crowded, they will swarm—that is, they will leave the hive as a group. They won't sting, and you may have to track them down and coax them into the hive. Splitting the hive and adding sections prevents this.

Fall: On a warm sunny day in the afternoon, take out the honey. Leave at least

▲ A swarm of bees gathered under the eaves of a house.

▼ Scraping comb off of a frame.

50–100 pounds for them to eat during the winter, depending on how long your winter is.

Winter: Keep the hive very well ventilated and protect it from wind. Check the bees' food supply and add sugar or sugar water to keep them from starving.

ZONE 2: ORCHARD

I am not bound for any public place, but for ground of my own where I have planted vines and orchard trees, and in the heat of the day climbed up into the healing shadow of the woods.

~ Wendell Berry

The Perimeter

Even though rigid lines represent the zones on your map, the edges should be very blurry in real life. At the edge of your

Zone 1 gardens is the orchard, but first you have to prepare the soil and develop nitrogen-fixing leguminous plants, such as clover or some species of shrubs. Then you can plant your orchard trees amongst the shrubs and small plants. These really shouldn't be in rows if they are for your own use, but if you are using the orchard to generate income, then you should plant in rows, always making sure to form them along contours. The edge between the two zones should be curved or zigzagged, as always.

Choosing the Right Trees

When mature, will the tree:
- Be shaped like an umbrella, or will it be more open? Open trees let in light for intercropping.
- Be tolerant to shade? If you want to grow smaller trees under your larger trees, pick ones that are tolerant to shade.

- Grow too tall? A tall tree grows very wide and will shade out undergrowth unless you want to spend the time pruning it back to an extreme degree.
- Need lots of water? Keep trees that don't need water away from trees that do, to simplify the watering process.
- Do well with what's around it? Some trees may stop other trees from growing well. Also make sure to plant male and female species together for pollination.

You must also choose species that do well in your region. In a cold climate, apple, pear, quince, cherry, peach, plum, apricot, filbert, chestnut, walnut, hickory, olive, loquat, and pineapple guava may be grown. The species must be disease-resistant, which is more likely with a heritage variety.

Intercropping

Intercropping has been discussed previously in this book but is just as

▾ A well-pruned orchard.

important in the orchard as in other places. It is here in the orchard that the forest garden model has a chance to really play out. The species that you choose should be resistant to disease, won't compete with other plants for water and nutrients, and can function as a windbreak. Under the trees you can grow green manure, nitrogen-fixers, forage crops for chickens or sheep or a pig, repel insects and grass, grow flowers and herbs for bees, or grow vegetables until the trees get too big.

To stop grass from growing under the trees, a variety of small plants can be grown. These are really needed only in the first few years, when young trees are competing with the grass.

- Bulbs like daffodils and onion species come up in the spring and die off by summer.
- Dandelions and comfrey have deep spike roots and leaves that cover the ground.

- Fennel, dill, tansy, carrot, Queen Anne's lace, catnip, and daisy all attract wasps, bees, and friendly birds.
- Clover and leguminous plants cover the ground and create nitrogen in the soil.

Desert Orchards

When planning an orchard in a dry region, the first consideration is species. Choose trees that don't need much water or that can withstand drought. The trees will have to be spaced farther apart than they would be in a temperate zone, so that they won't compete for water, and it is a good idea to plant them during the rainy season. Trees should also be mulched. In deserts, rocks can be used to protect the roots from heat and damage and act as a thermal mass that keeps the roots warm at night. Palm leaves or brush can be propped over the tree to protect saplings from the sun, and fencing or

▾ A desert orchard.

The Ultimate Guide to Permaculture

dogs should keep nibbling animals away. Interplant leguminous species in between.

The most efficient way to water trees in dry climates is by either using drip irrigation, which is a pipe system underground, or by building roof water and storm drains that lead into swales. The trees can be planted on the edge of the swales to take advantage of the water. If the soil is very sandy, each tree can also be planted in a hole that has been lined with mud clay so that the sand won't collapse and water will be retained.

On a slope, trees should be planted in zigzags with logs or water runoff ditches running between them. At the top of the hill the hardiest trees can be planted, with progressively less tolerant species towards the bottom where the deepest soil and most water will be.

In a desert, you can be more flexible with your zones. If valleys and streams flow through the property, meandering through all the zones, you may need to plant trees along the pathway of the water or right next to the house to take advantage of graywater. It is much easier to grow a tree where there is water already rather than bring water uphill to a dry place.

Animals in the Orchard

While the orchard is getting established, you will not want to let animals in, or they would destroy the young saplings. But, once the trees and the bushes and herbs in the undergrowth are reasonably well grown, you can let very small chicken breeds in. The chickens will eat insects and the fruit that falls to the ground (stopping pests from gathering), fertilize the soil, and scavenge free food. This can be done at a ratio of about 100 chickens per acre (0.4 hectares). When the orchard is around 3 to 7 years old, you can let the pigs in, and they will do the same

▼ These pigs are aerating soil and clearing the land.

job. After 7 years, you can let sheep in, and after 15 years you can let cattle in.

Pruning

Pyramid: This type of pruning keeps the tree quite a bit smaller than other methods. It is a good idea to do this type of pruning after April rather than in the winter, to prevent silver leaf disease. It is a fairly straightforward strategy. You would carefully cut the tree into a pyramid shape.

1. A sapling in the first year should be cut back to a height of 2 feet (60 cm).
2. In the second year cut off 18 inches (45 cm) from the top of the main stem and trim off the ends of the remaining branches to just above the previous year's healthy bud.
3. The third year, cut 18 inches (45 cm) off the top of the main stem and trim back the ends of the top few branches

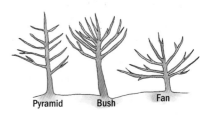

▲ Pyramid, bush, and fan pruning.

to just above the last healthy bud, or about 10 inches (25 cm).

Bush: Also called *open-centered*, the bush shape has a stem of 2 ½ feet (0.8 meters). The aim is to create a shape that has (obviously) an open center. This is done in early spring.

1. On a tree that is grown, pinch off any buds on the bottom of the trunk and pull off any *suckers* (shoots that grow off the roots and out of the ground).

Silver leaf disease

Found all over the world, *Chondrostereum purpureum* is a fungal disease that attacks just about any deciduous trees. Any home orchard is susceptible to it, and it can spread between species through wounds in the bark. It is recognizable by the silvery sheen it makes on the leaves when it damages the leaf cells. Apples tend to be fairly hardy and can usually recover, but other species can die from it. In the meantime it reduces the amount and quality of the fruit you do get. To prevent silver leaf, it is important to follow smart pruning strategies.

- Use sharp, high quality pruning shears so that the cuts you make are clean and smooth, and so the wood won't split. If the branch is big, use a saw to make a clean cut.
- Prune in the spring on a warm sunny day, not in the winter. Silver leaf prefers cool, wet conditions.
- If you are going to use wound dressing after the cut dries, do it on the same day. The dressing should be applied thickly with several coats. You should also know that pruning paint may be expensive, but making your own is ineffective. If you can't buy some, don't use any. The tree's natural defenses will help it heal, and homemade stuff will actually prevent it from healing.

---

2. This type of pruning is not very extensive. You should only have to remove stems and branches that are crossing, vertical, weak, or diseased.

3. If you still need to thin the branches a bit more, you can do so in July.

Fan: Fan training is used when a tree is grown up against a wall or fence at least 6 feet high and, hopefully, facing south or southwest for most varieties. Dwarf varieties work the best for this.

1. The tree should be planted 6 to 9 inches (15–22 cm) away from the wall, angled slightly towards it.

2. It is best to do the cutting in the spring, when branches can heal quickly. In the first year, when a sapling has no branches, cut back the main stem to 15 inches (38 cm), making sure that there are at least three strong buds.

3. In the summer, put two stakes into the ground at 45° on either side of the tree and tie the two side branches to them to start creating the fan shape.

4. Repeat the next year for two more inside branches at a lesser angle.

5. When a tree is already a couple of years old and hasn't been trained, you will have to cut back the main stem to about 15 inches (38 cm), put in the stakes, and cut each arm of the fan by two-thirds to just above an upward-facing bud.

6. At this point the tree will have two arms extending from each side. In the summer, tie four shoots from each arm 30° from the main arm so that they will create a fan shape.

7. Pinch off any shoots that are growing out towards the wall and all others back to one leaf.

8. The next spring cut back the four branches on each side by one-third, just above an upward-facing bud if you can.

▾ Small orchards provide hundreds of pounds of fruit per year.

DUCKS

There is one order of beauty which seems made to turn heads. It is a beauty like that of kittens, or very small downy ducks making gentle rippling noises with their soft bills, or babies just beginning to toddle.

~ T.S. Eliot

How They Fit into the System

Ducks are the gentlest and most versatile poultry. They eat algae and weeds from ponds, slugs, snails, grubs, soft greens and grasses, water plants, small tree greens, and grains and at the same time fertilize the water and the soil, improving fish production. They will walk on small plants, and they do eat some of them too, so they work better in a well-mulched area with plants that are well established. They also need less care and feeding than chickens, although they need more planning. While a few bantam chickens can be thrown in a greenhouse, a flock of ducks needs lots of water and grazing. If you have other animals that need to drink from watering troughs or ponds, the ducks must be separated from those water sources, or they will make them too dirty. The best system is an enclosed area just for ducks around a small pond with an island in the middle for them to nest on, usually in or on the edge of Zone 2. Around the pond could

be a forage garden, and if you have fish, the duck's manure will feed algae-eating organisms and help grow pond plants. You can keep 25 ducks per acre of pond surface. If this pond area is next to the Zone 1 garden, you can open it up now and then so that they can eat the slugs and pests, but only when the plants are at least as big as the ducks.

Duck Care

If the ducks have adequate water, a grassy yard with new grass, and a forage garden with bugs in it, then you won't need much extra feed. Ducks need young grass to eat, and if their pasture is too small and unvaried, they will quickly destroy a grassy backyard. If you must give them additional food, wheat is the best grain for ducks and goes well with oats. Hard round fruits and vegetables need to be crushed for them first. Liquid milk and hard-boiled eggs are good sources of protein for laying birds, and all ducks need calcium from eggshells or seashells, and grit. Ducks, unlike other poultry, need a little more niacin in their diet, but lots of fresh greens or peas should be enough to provide them with what they

need. In turn, they will give you eggs, meat, feathers, pest control, and fertilizer.

<table>
<tr><td colspan="2">

Homemade poultry chick feed

30% grain: finely ground wheat, a little corn, and oats

20% protein: fish meal, meat meal, yogurt, cottage cheese, worms, bugs, and grubs

50% greens: alfalfa meal, alfalfa leaves or fresh greens such as chopped lettuce

Extras: wheat germ, sunflower seeds, linseed meal

Sand for grit

Ground shells: mussel, snail, oyster, or eggshells

Homemade poultry adult feed:

30% grain: finely ground wheat, a little corn, and oats

15% protein: fish meal, meat meal, yogurt, cottage cheese, worms, bugs, and grubs

55% greens: alfalfa meal, alfalfa leaves, or fresh greens such as chopped lettuce

Extras: wheat germ, sunflower seeds, linseed meal

Sand for grit

Ground shells: mussel, snail, oyster, or eggshells
</td></tr>
</table>

Ducks are social creatures and need a flock to be happy, so you'll need to have at least two ducks or more. Each duck needs four square feet of housing. During the day this can be a three-sided shelter near the pond, but at night ducks need to be kept away from predators. They can be put in the barn in a room with a bed of straw, or in a simple shed. This shed should have a door for people so that you can harvest all the valuable fertilized straw. Ducks can

usually fly and need to be clipped, or else they will fly around your whole property or even leave completely. Use big scissors to clip off the ends of long feathers of one wing when they first grow, and after each molting after that. Don't cut during the molting, or you may cause fatal bleeding.

In the winter they will run out of forage and can quickly turn the area around the pond into mud. Throw down another layer of mulch, such as fallen leaves or hay, as they stir up the earth. Ducks can tolerate freezing temperatures as long as they can still run back into their three-sided shelter away from the wind when they need to.

There are several breeds that are popular with duck owners because they are better for eggs, meat, or both. Khaki Campbell is the most popular and was the first domestic duck breed, followed by Indian Runners. They produce just as many eggs as chickens, but are not good meat birds. Meat breeds include Muscovy, Rouen, and Pekin. To find a breed that is *dual purpose* or works well for meat and eggs, you will have to look to history and pick a heritage breed. Heritage breeds are types that were raised by small farmers over hundreds of years and are now not as commercially viable. Because of this, they are dying out. Ancona, Appleyard, Buff, Magpie, and Saxony are good dual-purpose breeds. Saxony is probably the best of these for their foraging and egg laying ability.

Breeding

Duck males and females are difficult to tell apart. The female will have a loud, raspy quack, and the male will be a bit quieter or sometimes silent. Some male ducks will also become very protective of the females. If you have a motherly duck, it is best to let her raise her own ducklings,

▲ Mother duck roosting.

▲ Chickens can run in the garden at certain times of the year.

as she'll do a better job than you. Ducks will start laying in the spring, when they are around six to seven months old, and keep laying for three years or longer. They always lay in the morning and are very scheduled, so let them out of the barn after 10:00 a.m. and then lure them back into the barn in the evening with a handful of grain. A mother duck and her ducklings should be kept separate from the other ducks until the chicks are six to eight weeks old.

CHICKENS

People who count their chickens before they are hatched, act very wisely, because chickens run about so absurdly that it is impossible to count them accurately.

~ Oscar Wilde

How They Fit into the System

The goal is to feed chickens from Zone 2 and profit from them in Zone 1, and so they are located on the very edge of Zone 1 or as close as possible to it. They provide meat, eggs, feathers, fertilizer, pest control, and weed control. The chickens live in a coop, which has an attached pen that runs along the border of the Zone 1 gardens and the Zone 2 orchard. It should also have a second pen for chicks, surrounded by spiny shrubs to protect them from hawks. The trees and ground in their pen should be well mulched with straw, corn stalks, sawdust, yard waster, or bark, with wire mesh around the trees holding the mulch in. The chicken run can be planted with fruit trees (which will drop fruit on the ground), grains, corn, sunflowers, and greens. When weeding the garden, the waste can just be thrown over the fence into the chicken run. The run is divided into several pens planted in succession, so the chickens can be rotated when the plants are ready, and each pen also has a log on the ground. The log is left to sit for a while and can then be flipped over to reveal all the pill bugs and worms. The fence dividing these pens and keeping the chickens from your other gardens should be at least five feet high.

Chickens can't be let into the mulched gardens or orchards that are for your use, which are typically in Zones 1–3, because they'll destroy the mulch. However, you can let them into an unmulched orchard. When the orchard is young, only let bantams or very small chickens forage where they will eat dropped fruit and weeds and fertilize the soil. There should be no more than 50–100 chickens per acre in the orchard or in a pasture with other breeds. Twenty-five light breed hens producing eggs will eat a quarter pound of food per day, and a heavy breed even more. If you are using chickens as a cash crop, then you may grow 300 chickens per acre on a well-planned forage pasture if no other animals are with them.

Alternatively, and this is especially the case for urban dwellers, chickens can live in a *chicken tractor* or *ark*. This is simply a small house for fewer chickens that is built to be moved around a yard or pasture. There are a myriad of designs and even commercially available chicken tractors made out of plastic. The house itself usually doesn't hold more than 3–8 chickens, and the pen is much smaller, usually only twice the size of the coop. Each chicken needs 3 square feet (0.3 square meters), which includes space in the pen. A large pen for more birds is too heavy to move frequently, which is why they are so small. The floor is open, of course, so the birds can eat, and the poop falls down onto the ground.

The pasture and pens should have trees and shrubs such as pigeon pea, berries, and fruits, as well as acacias, clovers, grasses, chicory, comfrey, and dandelions. Insects and larvae can be introduced by having a large manure pile or mulches, or you can grow your own crickets. You can give them cottage cheese or other dairy for more protein. They also need ground shells for minerals and grit to help them digest their food, which can come from their own eggshells, or from mussels or snails that you raise. If you don't provide this, they might eat their own eggs. Gravel, sand, and pebbles can also be put in a container on the ground for grit. See the section on ducks for a homemade poultry feed formula.

Chicken Plants

These can be planted in your pens and pasture where chickens eat. See the chapter on plants for more information.

Alfalfa
Amaranth
Autumn olive
Barley
Buckwheat
Chickweed
Chicory
Clover
Comfrey
Corn
Cucumber
Currant
Dandelion
Elderberry
Fava beans
Fennel
Fruit and nut trees
Honey locust
Oak
Oats
Quinoa
Russian olive
Rye
Siberian pea shrub
Stinging nettle
Sunflower
Swiss chard
Vetch
Wheat

Chicken Coop

The coop must be at least 3 square feet (0.3 meters) per chicken. A manageable size is 7 x 7 feet (2.1 x 2.1 meters), or 49 square feet (4.5 square meters), which is enough space for 16 chickens. A small chick-house can be attached to the side that is fortified from predators. Unlike a traditional coop, a permaculture coop is set on stilts so that the manure can simply drop down, either into a swale or into a waiting wheelbarrow, through slats in the floor. Inside, you will need roosts and nesting boxes. Roosts are long poles or boards at least 18 inches (45 cm) from the wall and low enough for them to fly up to. Nesting boxes are square boxes around the wall about 18 inches (45 cm) from the ground, and at least 12 x 12 inches (30 x 30 cm) in size. These boxes can be attached to the outside of the coop with a lid so that you don't need to go in to disturb the chickens every day to get the eggs. The general rule is one box for every two hens. The chicken door should be 12 inches (30 cm) high and incorporated into the peak of the roof. Not only does the door serve as ventilation, but the height also deters predators. A ladder to the door opening should be 5 feet wide with rungs 3 feet apart—very difficult for foxes to climb, but the rungs are close enough for chickens to comfortably

Chicken coop with fox-proof door.

traverse. In a cold climate the coop needs extra insulation and should be able to be closed up at night. The gaps in the floor may be a problem as well, but unless your climate is extremely cold, adding a thick layer of hay as bedding may be enough.

Keeping chickens healthy
- Disinfect the coop and equipment before you bring in new chickens.
- Keep bedding clean and dry, cleaning it out once a week.
- Clean and disinfect the coop once a year.
- Provide fresh water every day, making sure it is always available.
- Separate old birds from young birds to prevent the spread of diseases.
- Provide a small amount of crushed shells every day.
- A grit supply and cleanliness prevents worms and parasites.
- Cannibalism is prevented by adequate space and nutrition.
- Birds need vitamin D. In a northern winter you may need to provide cod liver oil.

Chicken Tractor

The most common small design for a chicken tractor is an A-frame, with a triangular house and a totally enclosed triangular pen covered in chicken wire. The other common design is a rectangular pen. The house takes up 1/3 of the total space and has a square cut out for the chickens to go out into their tiny run. A side wall or roof of the house is hinged so it can be opened up for collecting eggs and cleaning.

▲ A-frame chicken tractor.

Another feature that makes moving the chickens around much easier is wheels. The wheels must be set very low to the ground, as any gaps around the bottom of the pen can allow predators to get in. The solution to this is an extra wood barrier set around the pen that is removed when you have to move the chickens. Others have made larger, full coops on wheels with a removable ramp. These do not have an attached pen and are instead rolled out into a pasture. At night the chickens must be shut in the coop.

Breeds

There are so many different types of chickens that when people start to shop for their first batch, they are often overwhelmed. In North America, most of our eggs come from white hens that lay white eggs, but there are hundreds of other breeds from which to choose. These are divided into two categories: *light* or *heavy*. The light breeds can fly short distances, aren't good mothers, and aren't as hardy as the heavy breeds, but they are excellent at foraging and don't usually need supplementary food. The heavy breeds don't fly, are better mothers, and usually lay brown eggs. They tend to be hardier and lay eggs longer during the season but aren't as good at foraging. Some breeds are also much nicer than others, cutting down on pecking order problems. When chickens

▲ Orpington hen.

see a weakness in a bird either because of size or injury, they often gang up on it and peck it repeatedly, sometimes to death. Orpington is one of the most popular dual-purpose breeds. These chickens are fairly laid back and easy to handle. They are a good choice for a beginner.

Eggs

Chickens *molt*, or lose their feathers, once a year and usually in the fall, and they also stop laying. The feathers will start falling from the neck, then the breast, thighs, and back, and then the wings and tail. Usually, molting is triggered by less sunlight from shorter days, but by using a light on a timer in the coop you can keep the egg production up. Molting in the summer may be caused by stress, such as a food or water shortage, disease, cold temperature, or sudden lighting changes. Chickens also stop laying just because they get old. Their comb, vents, and wattle become shrunken and pale, their body will be smaller, and as time goes on their vent, eye ring, and beak will become yellow. However, chickens are amazing egg providers and usually produce at least one egg a day, giving you about 300 eggs per year.

Breeding

In a permaculture system, ideally you would want your chickens to hatch their own eggs, or at least the ones that you don't eat. This means you need to keep a rooster (which you can't do in the city), and there are a few other drawbacks as well. Roosters are loud and aggressive.

You also need the right breed for your hens. Heritage varieties are more likely to get *broody*. A broody hen is one with the instinct to sit on a nest, which most hybrid modern breeds don't have. Even then many get a nest started and then forget all about it. To encourage her, you can build her a private nest out of wood—about 15 x 15 x 15 inches (38 x 38 x 38 cm)—with a roof. It sits right on the ground and could have a little straw in it. You can also use a *dummy egg,* which is a wooden egg that you put in the nest to trick the hen into thinking she needs to sit there. When the weather is warm she will lay one egg a day and start *setting* or incubating the eggs. Don't let anyone disturb her or the nest, or she may abandon it. After 21 days the chicks will hatch. Many people take the

hatched chicks into the house until all of the eggs have hatched because the hen often will be torn between sitting on the eggs and taking care of the new chicks— she could end up losing both. Put the new chicks into a brooder (see below for more information on brooders), and then slip them back under her the first night after hatching. At this time you can also clean up her nest and take out any eggshells and dummy eggs. You can add in a few orphan chicks if you want. The next day watch the hen. If she pecks any chicks, return them to the brooder. Normally a hen breaks up big seeds for her babies, but if she doesn't, use the same methods for brooder chicks to feed them. The family should be kept away from other chickens—other mothering hens can cause fights, and adult birds may peck and kill chicks.

A brooder is simply a box with a light that keeps chicks warm like a broody hen. You can buy a brooder, or you can make one by using a box with a red heat lamp and a thermometer. The bulb should be low-wattage, and the temperature should be about 95°F (35°C). A box 30 inches (76 cm) square with a 69-watt bulb can brood 50 chicks. Put your hand down to test the heat. If it is uncomfortably hot, the lamp is too close or is too many watts. Check the chicks 2–3 times per night the first week. If they are cold, they will huddle under the

light, and if they are too hot they will stay near the walls. If they are content, they will look and sound like it by cheeping happily. Decrease the heat 5° per week, so that in 6 weeks it is 70°F (21°C). Then you can turn the heat off unless it gets chilly. As the chicks get older, tape another box next to the first and cut a door. Hang heavy cloth in the door, and the other room will be cool, so the chicks can run in and out. The light should be red or green and very dim, as a sudden adjustment to total darkness could kill them. Chicks also need water and bedding. You will need 1 gallon (4 liters) of water per 50 chicks. Keep the water clean and full at all times and make sure it is room temperature, not cold, although you can't put the water right under the heat lamp or it will be too hot and evaporate. The first week use burlap or cloth rags (with no loose threads) laid out flat for bedding over a layer of newspaper. Then, when the chicks have learned what food is, graduate to a thick layer of black and white

shredded newspaper, hay (not straw), or wood shavings. If you do use wood shavings, the pieces should be too big to fit in a chick's mouth. Stir the litter every day and remove wet spots to prevent spraddle legs (legs turning outward) and infection.

If you hatched your own chickens, wait until they start pecking at the floor to give them food. If you bought them, have food ready because they will already be three days old. For the first week put the food in an egg carton or on a piece of cardboard so that the food is up to their eye level. Once they figure out what it is, put the food in a container that is difficult to walk in and scratch food out of but short enough to reach in. Each chick will need 1 inch (2.54 cm) of feeding space until they are 30 days old, then they will need 3 inches (7.6 cm). Don't put the feeder right under the heat lamp, and only fill it half full to stop them from throwing the food out. Clean out the old food each time you fill it, and keep it full most of the time. Chicks

▾ **Newly hatched chick.**

need small grit to help them digest food, which you can give by sprinkling sand on top of the feed. This last step is simple but absolutely necessary, especially if you are making your own feed rather than buying commercial feed.

When they are 4 weeks old, chicks can stand on anything, so if you have a chick house outside attached to your coop, transfer the brooder to it then. When they are 6 weeks old you can remove the box, and when they are 10–12 weeks you can put them with the other chickens (if they are big enough and the weather is warm). These small chickens are called pullets and should be moved before they start laying.

protect your property from predators and provide eggs, meat, and feathers. They should only be allowed into a well-established area so that they won't squash any young shoots, and they will eat fruit and vegetables so they should be removed before your garden ripens.

Geese can be let into the Zone 1 vegetable garden after plants like strawberries and tomatoes have grown to the point that they won't sustain any real damage when the geese walk on them, but usually they will live in Zone 2, especially if you have ducks. Put seven geese per acre in the field when they are over eight weeks old, and let them graze until spring when the sprouts come up. There should be a fence around the field at least three feet high so they won't get into any other gardens.

Geese can live with a small pond just like ducks, but heavy breeds won't breed unless they have more water. Six geese is the maximum population per acre of water surface. If there is enough of it, they can also live just off grass pasture, and unlike ducks they eat older grass. Geese are meat birds and do well as watchdogs or guards, although they are very quiet. If you keep a goose for a long time, it can become too big for you to handle and will become dangerous if you aren't handling it every day, but a meat goose doesn't get big enough to pose a threat, because you'll eat it first.

Chick Dust

Chick dust is a powder that comes from dry chick droppings. It can get into your lungs and over time can cause lung disease, which is a good reason to either get chicks out of the house quickly or raise them in a separate area altogether.

GEESE

**Oh, nature's noblest gift,
my grey goose quill,
Slave of my thoughts,
obedient to my will,
Torn from the parent bird
to form a pen,
That mighty instrument
of little men.**

~ Lord Byron

How Geese Fit into the System

Geese eat grass and weeds and in return will fertilize the soil while leaving your crops and mulch alone. They also

Goose Maintenance

There are breeds of geese for eggs and some for meat. Dark breeds are harder to feather when butchering, which is why you always see pictures of the traditional white goose in farm scenery. It is difficult to tell the difference between a male and a female unless you are practiced in flipping them upside down and looking in their vent. The easiest but less accurate way is to watch the flock for geese with a broader head, longer neck, and more aggressiveness. A *gander* (male) has a deeper, louder call, and is much more aggressive. It is fairly simple to figure out if you watch during mating. Don't eat a goose over three years old, which you can tell by the soft, yellow down on its legs.

The goose house needs to be 10 square feet (0.9 square meters) per goose, and their yard needs to be very roomy, 30–40 square feet (2.8–3.7 square meters) per goose. The house doesn't need to be very fancy, just a simple shed that is very dry. A box feeder can be located inside, but the water trough should be outside under an awning. The floor should be covered with clean bedding such as chopped straw.

Geese can eat the same food as the ducks, as long as it has the correct ratios with 15% protein. However, they are able to live entirely off pasture and, unlike ducks, don't mind eating older grass. They do mind eating alfalfa though and generally won't touch it unless they are very hungry. In the winter provide dried grass, hay (not alfalfa), corn fodder, grain, and whatever scraps you would feed ducks.

Geese need to be clipped, just like ducks, after each molting. Clip 5 inches (12 cm) off the feathers of one wing, being careful not to clip the wing itself or during molting, causing permanent injury or even death. Geese are big and unwieldy, and an irritated goose can bite your face. Always pick them up backwards with the head

facing towards your back and the wings pinned under your arm.

Gather the goose eggs twice a day. If you want to raise goslings, you will need a gander. A big gander can service 2–3 geese, and a smaller one can handle 4–5. Once he picks, he will stick with the same females every year and help raise the goslings. Keep the ganders separate and let them in with the hens in late fall or early winter. It is recommended to wait until a goose is two years old before breeding because the quality of its eggs is so much better. Geese will want to brood outside, in an old tire with straw or a tiny brooding house. When a hen starts to lay, leave two and take the rest until the nest is full. Until she has a full nest, she may not start to set, and you could lose some of the goslings anyway, so doing this will also provide you with goose eggs and allow you to keep the hen laying. It takes between 28 to 35 days for the eggs to hatch. Around day 20 you will have to spray the eggs down completely with warm water and turn them over yourself.

Goslings raised in a brooder need the same floor space as ducks, 1.5 square feet (0.14 square meters) until 7 weeks and then 2.5 square feet (0.23 square meters) after that, but there should only be 25 goslings per 250-watt heat lamp. They will need to be fed 4 times a day, with enough food to eat in 15 minutes. This includes tender, green grass or weeds, along with a bit of duck food and some grit. At 5 to 6 weeks they can survive completely on a big pasture (1 acre or 0.4 hectare per 20–40 geese), or you can add some grain. Goslings can be butchered before winter as long as their pinfeathers aren't growing in, which happens in cycles. The grease has traditionally been used for frying,

pastry, and hand salves, since goose tends to be greasier than other meats.

PIGS

I had rather be shut up in a very modest cottage with my books, my family and a few old friends, dining on simple bacon, and letting the world roll on as it liked, than to occupy the most splendid post, which any human power can give.

~ *Thomas Jefferson*

How Pigs Fit into the System

Pigs are very efficient foragers and will eat any fruit that falls on the ground (helping to deter pests), grass, herbs, vines, and nuts and will dig up roots with their noses. This rooting action makes them excellent natural plows, preparing and clearing a garden bed before planting. The best place is a shady, treed area full of waste material and weeds and not too muddy. Pigs can be allowed into the orchard when the trees are at least a few years old. They will also eat any food waste and will eat 25 pounds (11.3 kg) of food per day. The pasture should be prepared with a chisel plow, and lime should be added. It should be planted with legumes, comfrey, endive, and grass. There should be no more than 20 pigs per acre (0.4 hectare), which will clear the whole thing. They will remove blackberries and scrub, and after they clear the pasture, the pigs can be taken out, the area can be replanted, and cattle can be allowed in. After the cattle clear it, replant and let the pigs return. Some plants can be grown as forage as well, including cattail, legumes, chicory, comfrey, and duck potatoes. A hundred pigs penned on 5 acres

(2 hectares) will eat through 100 acres (40 hectares) in 18 months, which can be a good thing if you have lots of land to clear, but everyone else needs to be careful to keep the population low.

In a cold climate pigs will need a three-sided shed with a soft dry floor in the pasture, and for breeding a *farrowing* pen will be needed. Farrowing is when a sow gives birth to piglets. She will use it for three months out of the year, and so if you have two sows, they can take turns. A lactating sow needs 7 pounds (3.2 kg) of grain per day when she gives birth, working up to 12 pounds (5.4 kg) when the piglets are a few weeks old. Each sow should have at least 2.5 acres (1 hectare) for forage just for herself and the piglets. All pigs also need a large and reliable constant supply of water. Since pigs don't sweat, an automatic waterer that the pigs can turn on and use to spray themselves will keep them happy.

As you are probably raising them for bacon, you may have to feed some grain for a couple of weeks at the end to fatten them up. If you don't care about bacon and just want healthy pork, then forage will do.

Pig breeds are divided into dark and light breeds according to their skin color. For the small farmer interested in a breed that will raise its own young, white breeds are often chosen because they tend to have good mothering instincts and are excellent foragers. Chester White, Yorkshire, and Landrance are popular mothering breeds that produce large litters, but these aren't known for their foraging ability. Alternatively, you could choose a heritage breed that has both skills, such as British Saddleback, Large Black, or Tamworth.

Breeding

Sows are generally gentle if you handle them often, but boars are not. They can be aggressive and can injure you easily, so

you may not want to keep your own. If you do want to keep a boar, avoid the Chester White breed, which are worse. The sow will go into heat if she isn't nursing, and you can bring them together then. It takes four months for her to give birth, which she will do in your farrowing pen. The piglets need to be kept warm, about 86°F (30°C), and they will huddle together even then. The farrowing pen, which is simply a well-ventilated shed with a door that is big enough for the sow to go in and turn around in, has two dividers on each side for the piglets to be able walk through so that the sow won't accidently lay down on one and smother it. You will also need a *creep feeder*. When the piglets are around a month old, they can begin to be weaned, and a feeder on one end of their grazing area that only the piglets can get into will make this process easier.

▼ Farrowing pen.

Market weight for pigs is around 250 pounds (113 kg), which yields 140 pounds (63.5 kg) of meat. This takes about six months, depending on how much grain you are feeding them. If it is getting close to winter and the pigs haven't put on that much weight, then forage hasn't been enough, and you'll have to supplement it with grain. Butchering pigs is very labor-intensive and requires some extra sanitary procedures than with other animals, and so it is recommended that you delegate that task out to a professional. Unless you are raising

The Ultimate Guide to Permaculture

them commercially and have an interest in setting up a professional butchering facility, it is worth paying someone else for the amount of time spent butchering one hog.

GOATS

He who lets the goat be laid on his shoulders is soon after forced to carry the cow.

~ Italian Proverb

How They Fit into the System

Goats are exceptionally good at clearing pasture and effectively clear the toughest brambles and unwanted vegetation. They can be temporarily used for this purpose by penning them or tying them with a halter and moving them from place to place. Goats can be so destructive, however, that it is only recommended to keep a few for milk and meat production. More than one goat per person in your family is unnecessary. If you do use lactating goats to clear a pasture, you will have to give them a little bit of grain to keep their milk production up.

Goat Management

Goats are clever and can also jump high. The fence should be at least 4.5 feet (1.4 meters) high, with ¼ acre (0.1 hectare) per goat. Wrap trees with chicken wire so they can't strip the bark off, and make sure the fence does not have a gap wider than 8 inches (20 cm). If the goats can't see through it, they won't try to get out, but if you must, have a rail fence the make sure they can't squeeze through. Goats can unlock most standard latches with their tongue, so a padlock may be necessary. If the goat does try to get out all the time, put a Y shaped yoke on its head so it can't fit;

▲ **Goats will strip a tree of everything they can reach.**

soon it will give up trying, and then you can remove the yoke. The goat house can be any kind of sturdy three-sided shed in their pasture, or in the barn as long as there is 36 square feet (3.3 square meters) per goat, with clean hay for bedding.

Goats are very curious, so pretend to examine something while showing them some grain. Not only will a goat want the grain, she will want to see what you are looking at. You will need a second person to help you, and when the goat walks closer to investigate, the helper can grab her. If you spend a lot of time with your goats, this may not be necessary, as they can become quite attached to you. To find out the sex of the goat, a boy pees from the middle of his belly, and a girl has to pee squatting backwards a bit.

Goats will try to eat anything, including poisonous plants. Check their pasture for these common dangers:

- Milkweed
- Nightshade plants
- Buckthorn
- Cowbane
- Dog's mercury
- Foxglove
- Greater celandine
- Hemlock trees
- Henbane
- Ragwort
- Rhododendron
- Rhubarb leaves
- Spindle
- Water dropwort
- Yew
- Iris
- Azalea
- Beet leaves
- Evergreen trees

Each goat needs about 4 to 5 pounds (1.8 to 2.3 kg) of hay per day of mixed grass and legumes, such as alfalfa. That's about half a bale of hay per 10 goats, twice a day, placed on a hayrack. Since you probably have them on an overgrown pasture, they will eat less hay. Goats also need salt and water at all times. Goats won't lick a salt block, and so you will have to provide loose mineralized salt. While goats are destructive and curious and tend to get out of their fencing, in other ways they are very easy to care for. If they get lice, rub them down with vinegar. Trim the hooves once a month using a knife or hoof nippers, or they will keep growing. Goats are very susceptible to worms and under regular standards would be wormed three times a year. For organic standards this is not possible. The goats themselves will control their own worms in the way that they eat, by eating higher leaves first and working their way down, and wandering far distances over their pasture, which is the tree method

of pasture, is particularly valuable for them. They should be rotated to new pasture every three weeks, which is the lifespan of a stomach worm. They need to be kept in very clean conditions with fresh clean water readily available at all times. Unless you have lots of room for the goats to roam and lots of forage available off the ground, you may have to worm them.

Breeding

Start feeding the breeding does a quarter pound of grain per day. Start on October 1st and increase a quarter pound (0.11 kg) per week until the beginning of November when each doe is getting 1 pound (0.5 kg) of grain per day. This will increase the chances of having twins and triplets. For healthy older does, taper off this grain starting December 15 (six weeks after breeding), and then bring it back up again starting February 15 (six weeks before kidding). First-time breeders and unhealthy animals can keep eating grain all the way through. Bucks that are running in with the does will have access to grain as well, but it should be gradually tapered off when a buck is in good condition.

An average doe produces about 3 quarts of milk per day. About one 1.5 quarts goes to her kid, leaving only 1.5 quarts per day for you. If she is a new mom, she will give even less than that. Check the mother's udder twice a day after kidding to make sure it isn't too full. Her milk will come in three to five days after kidding, so during that time don't give her grain, or the milk will come in too fast. What you feed your milk goat can change how the milk tastes. If the milk tastes bitter, try removing high-odor foods from her diet, like garlic or cabbage. Goat milk may taste naturally

"goaty," but you can fix this by putting a pan of baking soda in her feed trough. Keep it full of soda, and in a few days the milk will taste sweet. In fact, any goat eating grain should eat baking soda to avoid a sickness resulting from too much acid. Pregnant does who are producing only a little milk should stop milking two months before kidding, but you can keep heavy-producing does milking all the way through continuously. Pregnant milkers need high quality feed, but not too fattening.

To have milk from a doe, she will have to breed with a buck. Whether you raise a buck yourself or buy an adult goat, always treat it gently and carefully so that the butting instinct is not awakened. Don't make a pet out of it and don't allow it to think you are a part of its herd. The best strategy is to teach him to lead and stand quietly for grooming and hoof trimming, and then leave him alone in his own separate pasture.

Breed your does 149 days or 5 months before you want to have kids. For small farms, having kids about April 1st is ideal, so breed on November 1st. Don't breed does that are less than 70 pounds or 2 years old because they will have health problems. One buck can service 50 does, but 30 are easier to handle. A buck should have his own sturdy pen, and as the does goes into heat, she can be put in with him until they breed. Then you will know the exact due date because does almost always kid exactly 149 days later. Does are in heat when they spend time sniffing and wagging tails towards the buck pen, and make more noise.

When a doe is close to kidding, you should check on her every morning, and if she has delivered, you will need to find the kids as soon as possible. It's best to be on

▲ Nanny goat with newborn kids.

the safe side and keep the goats nearby. A barn floor with dry bedding is ideal if they are familiar with the place. Give the doe clean, fresh water in a small bucket and lots of good hay. When she starts to give birth, you can clean the nose and mouth to help the kid breath. Otherwise, leave her alone unless:

- her water breaks and 2 hours pass without seeing any part of a kid
- if she's in great pain and 30 minutes pass with nothing happening
- if she's totally exhausted and 15 minutes pass with nothing happening

To assist in the birth of a head first kid, pull gently downward with each contraction. If one of the situations above occurs or you can see a bum or neck coming out first, scrub your hands and arms well and make sure your nails are very clean. Reach in very carefully, determine the kid's position, and gently reposition it until its feet and nose are coming out together. After delivery, give the doe a bucket of warm water with some molasses in it and let her deliver the afterbirth. She may eat it, or you can bury it.

Wipe the kid's face, and if the goat isn't doing a good job of drying the kid, then go ahead and dry it off. If you find a kid outside after birth, bring it inside and wrap it up next to a heat source until it is warm. It isn't necessary to tie the cord, just dip the end in iodine or alcohol to kill bacteria. It's common for a goat to have two or three kids, and they should be standing up right away. If one doesn't, nurse within the first

15 minutes, helping it by holding the teat in its mouth so it can suck. If it doesn't suck, squirt milk in its mouth and then try to get it to suck again in 3 or 4 hours, or when it seems hungry. Some kids need help like this for 3 days. If it still doesn't suck after several hours, then you'll have to bottle feed, but only a couple of times so it learns to suck—then try having it suck on the mother. Don't keep the kid in the house or away from the mother for more than 6 hours, or the mother will reject it. For 3 to 5 days after birth the kids and mother should be kept separate from the herd until they are strong and nursing well.

A rejected kid must still be fed colostrum. Give the mother a little grain and milk her. Save the colostrum, put it in a bottle with a lamb nipple on the end, and feed it to the baby. If the kid is too weak to use the lamb nipple, use a human baby's nasal aspirator (or syringe). Never feed a kid cold milk. It is so important to get the colostrum into the baby; it is the difference between life and death. The kid will have to be fed every 2 ounces (57 grams) ever 2 hours on day one, gradually increasing to 3 ounces (85 grams) every 3 hours on day three, and 6 ounces (170 grams) every 4 hours on day seven. Work your way up to 8 ounces (226 grams) morning, noon, and night for 2 weeks, and then gradually to 16 ounces (half a kilogram) morning and night after you milk. If you can't give the kid goat's milk, the next best is cow's milk, but the kid may still get scours (or diarrhea) and die. You will have to butcher the kid if you don't have goat milk or if you need the milk yourself. It is possible to overfeed a kid. Normally a mother goat lets the kid start eating and then walks off after less than a minute. Frequent, small meals are better. After the kids are 2 or 3 months old, they can be weaned, but a bottle-fed kid will be more attached to you and become annoying.

Let the kids stay with their mother until they are strong and eating solid food well. This usually takes 2 months for singles and twins and 3 months for triplets. Then separate the kids from the mother at night and milk her in the morning, and then put her again back with the kids. In a few days they will learn the routine, although they will make a lot of noise about it. As the kids wean, keep milking her more and more so she will get used to holding more milk in her udder at a time and her teats will lengthen. You can either keep letting the mother feed the kids at night and only milk once a day, or you can gradually work toward milking her every 12 hours. If you let the mother nurse once a day, she will eventually wean them herself.

Besides the rejected kids that you can't feed, it is also likely that you will end up with more billy goats than you need. Nanny goats are valuable and you will hardly ever need to kill them because you can sell them instead, but billies don't give milk, and if you have too many, then you get no milk for yourself. All of the extras can be raised for meat until the fall. If you have them butchered in November, then you won't have to feed them over the winter.

Milking

Stainless steel equipment without seams are the best and also the most expensive containers for milk. Food grade plastic and glass will work also if they are seamless, but don't reuse plastic milk jugs from the store. You'll have to purchase containers. Rinse buckets, containers, and utensils in lukewarm water right after you use them, or they will collect milk deposits that are difficult to remove. Wash them after every use by scrubbing thoroughly in warm, soapy water, then rinse again in

▲ Acceptable seamless milk containers.

scalding water. Air dry upside down. Any cloth used for straining should be rinsed and boiled directly after.

Goat milking troubleshooting

The milk won't let down: Massage the udder either with the cleaning cloth or with a bag balm, or gently pat the udder like a kid butting her. If she still won't let down, real goatherds suck on the teat a little.

Drying up milk: This is done to let her have another kid. When you milk, leave a little milk in the udder. When the doe's milk production has reduced, milk her only once a day. Be aware of the signs of mastitis, as this is when she is most at risk.

Mastitis: The first sign is milk with strange texture: flakes, lumps, or strings. Don't drink it and don't throw it somewhere that an animal can lick it! Wash your hands very well after touching the animal because it is infectious. Even healthy does should be checked at least once a week by squirting the milk into a cloth. Feel the udder for tumors, large hard areas, or an *abscess*. An abscess is a red, tender swelling of a whole side of the udder, which makes it warmer and more difficult to get milk from. If you allow it to get worse, the milk may turn yellow or even brown or pink from pus and blood. The only cure is antibiotics. A bruised udder, getting too full for too long, or having had a previous case of mastitis makes her more susceptible.

Self-sucking: Goats are flexible enough to suck their own milk. Other than butchering, you can use an Elizabethan collar or side-stick harness that can prevent a doe from reaching her udder.

1. Clean the milking utensils in warm soapy water.
2. Put the goat in a *stanchion*, a frame that holds the goat by its neck. Many people have a milking stand that raises the goat a little higher for milking comfort. Brush the fur to get out loose hair and dirt, put down fresh bedding, and keep the long hair under the udder clipped.
3. Put some feed in the stanchion's trough. Milking is done every 12 hours, starting early morning before goats go eat. Always be on time, or the goat will get too full, which is painful. If you aren't a morning person, just make sure that you stick to a schedule.
4. Brush the doe and look it over for problems. Then wash your hands and dry them, and fill a bucket with water 120–130°F (49–54°C).
5. Wash her udder and teats. This helps the milk let down and removes dirt and bacteria, making better milk and a healthier goat. Wait a minute after washing to start milking.
6. The goat needs to be happy and relaxed for the milk to let down. Put your thumb and forefinger around the teat near the top of the udder, pushing up slightly and allow the teat to fill with milk. Then close your hand around it and squeeze the milk out while pulling down. Keep your hand away from the nipple hole or the milk will go all over. Squirt the first three squeezes into the ground because the first milk has more bacteria in it. Make sure you completely empty the udder or she will produce less and less until it's gone.
7. Strain the milk. Use a regular kitchen strainer lined with several layers of clean fabric such as a dishcloth, muslin, or even a cloth diaper.
8. If you choose to pasteurize, get a double boiler. Use a thermometer to heat the milk to 161°F (72°C), stirring constantly. Maintain that temperature for 20 seconds, then quickly remove

▼ **A healthy udder.**

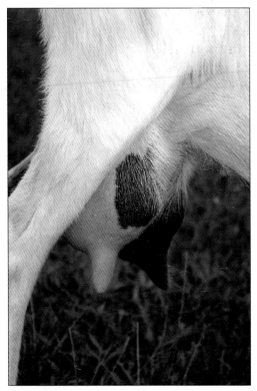

the milk from heat and immerse the pot in very cold water, stirring constantly until it gets down to 60°F (16°C).

10. If you don't pasteurize, put the storage container inside another container full of cold water. It needs to be chilled to 40°F (4°C) within 1 hour. Store any milk in the coldest part of the fridge.

SHEEP

In order to be an immaculate member of a flock of sheep, one must above all be a sheep oneself.

~ Albert Einstein

How They Fit into the System

Sheep can be let into the orchard to forage after the trees are at least 7 years old, but sheep must be carefully controlled. They must be taken out if they start to damage the trees. However, just because the orchard is not available doesn't mean trees can't be developed as a mainstay of their food within their grazing area. Trees provide food, protect from the elements, benefit the soil, and prevent erosion. Sheep also need a salt lick and a variety of other food types. A legume-grass mix pasture will feed 5 ewes and 8 lambs per acre in a northern region, but it needs to be rotated every week. In the last month of pregnancy, ewes need 0.5 to 1 pound (0.22 to 0.5 kg) of grain per day. After lambing, sheep with one lamb need 1 pound (0.5 kg) of grain, and sheep with 2 lambs need 1.5 to 2 pounds (0.7 to 0.9 kg). This amount can be gradually tapered off in the next 2 months. A ram needs 1 pound of grain each day during breeding season. This grain supplement can be offset by tree forage. Each sheep also needs 1.5 gallons (6 liters) of fresh, clean water every day. Every spring you will need to gradually introduce the sheep to pasture so they can adjust. In the winter they can forage from a harvested corn field and roots left in the garden. Each sheep will eat 75 pounds (34 kg) of grain and 10 bales of hay over the winter.

▲ Great Pyrenees dogs live with the sheep.

Sheep Management

Sheep don't need much housing. A three-sided shelter in the pasture is sufficient, along with a place in the barn for lambing or extreme weather. Sheep are very vulnerable to stress. Moving them in a truck, abrupt feed changes, or loud storms are enough to make them stop eating and even have a heart attack. They are also an easy target for predators. Stray dogs can "chase a sheep to death" without biting it, and of course any larger animal can eat a sheep. A dog is the most effective prevention. A border collie will herd the sheep, or you can have a guard dog live with the sheep in the field. The Great Pyrenees was bred for this purpose: to stay in the field and attack sheep predators.

Breeding

Breed your sheep in October, and in 5 months you will have lambs. If your sheep have hair on their faces, clip the wool away from the eyes, and also *tag* them, or clip the wool away from the vagina. Every time you do something to a ewe is a good time to check her feet. You can buy a yearling ram every year to breed with the ewes and then eat him afterwards, or you can keep a ram full time. Obviously it's a bad idea to use rams from the same family for breeding. Lambing is almost exactly the same as kidding for goats, and orphan lambs can drink goat milk as a replacement.

CATTLE

All the really good ideas I ever had came to me while I was milking a cow.
~ Grant Wood

How They Fit into the System

Cows can be allowed to graze in the orchard after at least 7 years if the sheep do well and aren't harming the trees. They should not be allowed to clear the orchard of vegetation but simply graze now and then to control the grass. Cows have four stomachs and are built to eat with their

The Ultimate Guide to Permaculture

heads down low, but even so they can still be destructive on the orchard trees. They need fresh, clean water at all times, and a salt block that is sheltered from the weather. The best dairy cow pasture is lush clover and alfalfa, with grasses and herbs like lavender, mustard, rosemary, and sage. In the winter they need tender green hay that was cut just after it bloomed, 2 to 3 pounds (0.9 to 1.3 kg) per 100 pounds (45 kg) of body weight. Every cow needs about two tons of hay over the winter. Don't give her more than she can eat until the next milking or she won't eat the old stuff. If you have only good quality hay or grass, she will still give milk, but if the quality drops, so will her milk supply. If you have no fenced pasture, you can stake a milk cow with a large bucket of water and move her three times a day, which helps utilize more pasture. After a cow gives birth, you can supplement with grains in order to increase her milk production, slowly increasing it as her milk supply grows and tapering it off again near the end of her lactation. The most she will ever need is 1 pound (0.5 kg) of grain per 3 pounds (1.4 kg) of milk she makes. With quality grass and 0.5 pounds (0.23 kg) of grain she will give about 90% of her milk-producing capacity. Cornstalks, washed and sliced root vegetables, and whole sunflower heads make great supplements. If you feed her with grains, make sure they are chopped or ground, and it is better to mix different types of grains together. Quality alfalfa pasture or hay may be rich enough that you wouldn't need a supplement.

Cattle Management

There are hundreds of breeds of cows, and they are divided into two groups: beef and dairy. Dairy cows make more milk and are gentler. There are a few breeds that have been bred to do both, such as Shorthorn and Brown Swiss. For the small farmer, dairy or dual-purpose breeds are the most versatile as the young bulls can still be used for beef, and the steers could be used as oxen. If you are inexperienced

with cows, buy a 4 or 5-year-old cow instead of a heifer because she will be more experienced in giving birth.

Cattle are red-green color blind and have nearly 360° vision, so they can startle quite easily. If something jumps into their peripheral vision suddenly they can freak out. Put blinders on and they calm down—what they can't see doesn't bother them. Their lives are based on habit, so they sometimes seem stupid because they are driven by routine, but they are actually quite intelligent. Be very patient, quiet, and gentle, and utilize their herd instinct—if you get one cow to do something, they all will follow. If you use food as a reward, you can call them, and they will come without any work on your part. Cows with calves are very protective, and cows in labor can be violent, so you might want to tie those for safety. Be careful around cows in heat and never *ever* turn your back on a bull. Never let your kids around a bull, and if a bull threatens you, eat it. Don't breed a bull younger than 18 months old.

Common Cow Problems

Downed cow: Sometimes a cow can't get up. It could just be a matter of helping her reposition her legs, in which case just get several people to pull her legs out on a level surface. If it is an injury, make sure she is fed and watered until she can be moved.

Scours: Scours is severe diarrhea caused by nasty things that attack a young calf's gut. Give the calf Kaopectate, which is a mixture of pectin and kaolin clay. The only prevention is providing a clean, dry birthing area and making sure calves get colostrum.

Heel fly: Heel flies lays eggs in the heel of a cow and then they hatch and crawl through the cow's blood vessels to bore holes through the skin and fly away. The only prevention is reducing the fly population with friendly predators like frogs or lizards.

Breeding

A cow can have a calf every year and provide you with milk almost the entire year, until she is 10–16 years old. Some cows have been known to have calves until 20 years of age, but that is rare. Some of the best dairy breeds are Ayrshire, Brown Swiss, Guernsey, Holstein, Jersey, Red Poll, and Devon. Don't breed a cow until after she is a year old, and be careful, because a three-month old can conceive. You wouldn't want her brothers to do the deed. You don't need to dry up a cow to get her pregnant again, but to have a healthy pregnancy she needs to put on weight, which she can't do if she is lactating the whole time. Breed her, and then dry her up 2 to 3 months before she calves. Her gestation period is 285 days (9 ½ months). Usually she will go into heat for 18–24 days, but it will be 30–60 days after she calves, so keep an eye on

her. Then you can breed her immediately. If you breed in the middle of May, she will have her calves at the beginning of March. A cow in heat will be restless, stamp her feet, and twitch her tail excessively, and she may make lots of noise. Her vulva may get red and swollen, and other cows will try to mount her, even females. A normally peaceful cow might suddenly run through a fence, or bulls might struggle to get to her. She might even try to mount you.

When your cow is close to calving, bring her into the barn and put down a thick layer of clean bedding. Check her frequently, but she should be able to calve alone. You can call a vet if she doesn't give birth after 3 to 4 hours of active labor. Make sure that the calf can get up and get milk (or colostrum) fairly soon after birth. If the mother doesn't clean the calf very well, it is fine to rub it dry, as there is not much

danger of her rejecting her calf. After you let them back in the pasture, the cow may try to hide her calf somewhere and just go to feed it during the day. Make sure you check on it, and if you want to move it, just pick it up and the cow will follow. If you find the calf and it has become chilled, bring it in the house, wrap it up, and warm it by the woodstove.

Milking

Keep your cow milked at least once a day or she will have too much milk and might get mastitis. You will get at least a gallon of milk or more per milking. The first 4 days after she gives birth is all colostrum, which needs to go to the calf. Once the colostrum taste is out, you can drink it too—you will have plenty of milk for both of you. A calf isn't often orphaned or rejected, so the easiest way to feed it is by letting it run with its mother until weaned. A family shouldn't need more than a gallon of milk per day for its own needs, and a couple of extra cows will provide enough to make a small profit. The care and milking of cows is exactly the same as it is for goats. When it is time to wean, wait for the calf to start being interested in other foods by offering grass and hay or by *creep feeding*. A creep feeder is a little structure with tempting foods in it that only the calves can get into. It is put into the pasture so that the calves get curious and nibble from it. As the calf eats more and more grass, allow him access to his mother less and less until he is only eating hay and grass. This is a good time to separate your young bulls from your heifers so that no inbreeding occurs. They should be weaned by 4 or 5 months old. During summer they will need at least an acre of good pasture. In winter a weaned calf will eat about 2 pounds (0.9 kg) of hay per 100 pounds (45 kg) of body weight per day. Three pounds (1.4 kg) of cornstalks and cobs correspond to one pound (0.5 kg) of hay.

▾ **An older calf in the process of weaning.**

The Ultimate Guide to Permaculture

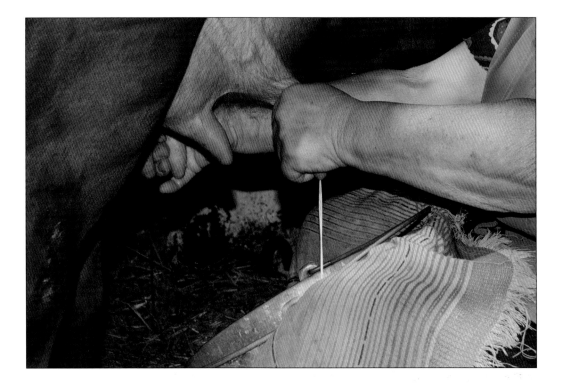

ZONE 3: BARN

A little and a little, collected together, becomes a great deal; the heap in the barn consists of single grains, and drop and drop makes an inundation.

~ Arabian proverb

Barn Design

Having a Zone 3 on your property is a privilege, but along with it comes the responsibility of building and maintaining a barn. The barn is the center of your extended zone activities, ranging from grain and cash crop harvest to animal shelter and food storage. It should be designed to save work. For example, if you build the barn on a slope uphill from the garden, a chute can be built to push manure under the barn for easy access, which can be rolled downhill easily.

- Orient the barn at a 45° angle to the prevailing wind so that the barn won't act as a wind tunnel. If this is not possible, build entrances from two different directions.
- Put the barn close to the manure pile and opening into a pasture. Water and feed should be nearby, and it should be easy to remove manure from all the stalls.
- Window vents should be placed high on the outside walls so that the animals can't reach them.
- Use natural lighting as much as possible, and make sure each stall has its own light bulb with its own switch.
- A horse stall needs to be at least 12 x 12 feet (4 x 4 meters) and 10 feet (3 meters) high. The bigger the stall the easier it is to clean. Walls into the center aisle can have bars for

ventilation but between stalls should be solid to provide privacy.

- Stall doors should have wood lower section and bars above so horses can see out.
- Make one stall a wash stall with a drain in the center of the floor (a slope of 4° going into the drain) and supply a hose for bathing. Storage, a sink, and counters are also nice to have.
- The feed room should be rodent-proofed with sheet metal. Mucking equipment should be kept somewhere other than the feed room, to prevent contamination.

A barn designed for permaculture can be a little different. The barn can have two levels and is most ideally built on a slope so that animals can go into the main floor upstairs without needing a ramp. Behind each stall are slats in the floor so that manure can be pushed through to fall down to the bottom level. This bottom level collects manure, which can be left there to decompose for a while or taken to the compost heap. In a very warm climate the bottom level doesn't have to be enclosed at all.

Flooring

The animals that you house in the barn all have a variety of different needs, and while the roof is important for keeping them dry, it is the floor that is the crucial component of your animal's well-being. Horses are particularly demanding in their requirements because they need a ventilated floor to prevent foot and health problems. The floor must be absorbent so that it does not become slippery, easily cleaned, and resistant to degradation due to pawing. Before you even build the floor, remove the topsoil to prevent settling, because any dips or divots in the floor can cause a problem.

Clay: Clay is an inexpensive option but is also high maintenance. It stays warmer than other flooring materials, but it can become slippery, holes can form, and it is difficult to clean. It should be placed

▾ The iconic barn, set at the edge of a Zone 2 garden.

▲ Two-story small barns for easy manure collection.

over a well-drained subfloor of crushed rock or gravel. If you mix sand with the clay (combine two parts clay and one part sand), it will be easier to maintain and have better drainage. Pack it down well and make sure it is level.

Sand: Sand is cheap because you won't have to add any extra bedding such as straw, but you will still have to change the sand regularly. It may not be the best option for horses, which may eat some of it and get colic.

Limestone dust: Limestone dust is a long-lasting floor that packs down as hard as concrete. It is often used in horse barns to fill holes because it helps control bacteria. It should be placed over a base of seven inches of sand, watered down, and packed hard and level. Lots of bedding is important to have since the floor is so hard.

Wood: Rough cut treated hardwood at least 2 inches (5 cm) thick is low maintenance but can be slippery. Boards should be placed over a base of 6 to 8

inches (15 to 20 cm) of sand or gravel with space left between large planks for drainage. Pack the cracks with gravel or clay.

Concrete/asphalt: Concrete is easy to clean and maintain, but it has absolutely no drainage. You will have to use lots of bedding. Horses kept on concrete or asphalt need to be outside for at least 4 hours per day.

Rubber floor mats: Rubber mats are the luxury product that horse owners recommend. They are pricey but easy to clean and add additional softness. They should be level, pushed tightly together, and installed up the stall walls as well. You will still have to use bedding, but the mats do help.

Sawdust: Sawdust is cheap and plentiful and works well, especially if you have wood chips or rubber mats underneath it. If you have horses, never use walnut sawdust or even sawdust milled right after walnut was milled, as it is deadly to horses. It shouldn't be too fine, or it can cause lung problems.

GRAIN

With coarse rice to eat, with water to drink, and my bent arm for a pillow—I have still joy in the midst of all these things.

~ *Confucius*

The Rice Paddy

It has been emphasized throughout this book that soil should not be turned or disturbed much at all. Grain is a staple of most people's diets, and yet how do we grow it without destroying the soil? Modern grain farming uses monstrous machines to turn and fertilize the earth, and even bigger machines to harvest perfect rows

▲ **Small rice paddy.**

of grain, and then the land is left to sit over the winter. This inefficiency can be easily solved with Masanobu Fukuoka's no-till method. Many sustainable farms follow a *rotational* planting schedule that involves planting legumes before and after the grain crop and includes a *fallow* period, a time when nothing is grown, to let the soil rest. The Fukuoka system, on the other hand, grows grain and legumes together continuously.

Not only is this system very sustainable, even more important is the low energy-input required. Not only is there no mechanization, the energy used in human labor is also very low. The farmer can eat an average 2,000-calorie per day diet and still produce 1,300 pounds (590 kg) of rice (or 22 bushels) on a quarter acre. Using animals for labor in the traditional manner

uses at least 5 times as many calories with the same results. Using a tractor uses at least 10 times as much.

The other benefit of this system is the small space required. Many self-sufficient homesteads and small farms avoid growing grains because they have seen large commercial grain fields and assume that it must take vast expanses of land to get any amount of grain. So they grow potatoes and keep a cow instead. But the reality is, to keep one human alive entirely on one food would take:

- 1,800 square feet of just grain
- 5,400 square feet of potatoes alone
- 13,500 square feet of just dairy farming
- 36,000 square feet of pigs alone
- 90,000 square feet of just beef

This is the greatest argument for a plant-based diet.

On the outer perimeter of your grain fields, a band of weed-control plants should be grown, such as comfrey, lemongrass, or citrus. These should be mulched with sawdust for even more protection. The grain strategy outlined here centers around rice and the construction of a paddy. If you can't have a paddy that fills with water, dry rice species exist that just need to be watered and have the additional benefit of being able to survive on monsoon rains alone.

First, level the ground and build a low mud retaining wall around the plot that can hold 2 inches (5 cm) of water. You may need to use a chisel plow the first year if the soil is extremely compacted. Spread lime or dolomite and a thin layer of chicken manure over the area and water it in. This soil disturbance and fertilizing needs to be done only once.

Cold Climates

In some areas it is too cold to grow rice, and you will have to use a system with shorter cycles. Spring wheat can be planted in the spring with oats or barley or wheat as the winter crop. You can also experiment with squash, melons, tomato, cotton, vetch, or sunflowers as a no-till crop.

No-Till Grain Strategy

This section should be more properly called the no-work method. The four principles of natural farming are no cultivation, no fertilizer, no weeding, and no pesticides. We already covered the *why* behind no-till methods when we talked about pioneer plants and the disturbance of the soil. It is important to understand here that we are not talking about soil that is never touched, but rather about soil that is aerated and loosened by natural means only. With this mindset we have to question weeding as well. What are the weeds doing for the soil? If the plant you want to grow is not being harmed in any way by the weed, then why pull it?

1. A variety of plants can be grown together. You may want to have several plots with different combinations, but each plot will always grow rice and white clover. Then you can add rye, barley, millet, winter wheat, or oats.

Type	Seed quantity per acre
Clover	1 pound (0.5 kg)
Grains	6–14 pounds (2.7–6.3 kg)
Rice	5–10 pounds (2.3–4.5 kg)

2. Rice seed is sown in early fall. It can be broadcast and covered with straw or made into seed balls. Seed balls can be made in two ways: You can mix the seeds with mud and press them through a wire mesh, or you can wet the seeds down and roll them in fine clay dust until they form a ball shape. This clay mixture can be made from a combination of potter's clay, compost, and sometimes paper mush.

3. In mid-fall, harvest last year's rice and lay it out on rice racks to dry for a couple of weeks. Thresh off the husks and straw and save it.

4. Within a month of the rice harvest, sow unhusked rice in the field and spread the husks and straw that you saved over it.

5. In the winter, if the rice has grown to 6 inches (15 cm), you can allow 40 ducks

per acre (0.4 hectares) to graze in the field. If you notice any spots that are growing in thin, plant more seed quickly. You don't want water accumulating during this time of year or it will freeze, so keep the rice drained.

6. In spring, check again for thin spots and sow more seed if you need to.

7. In late spring it is time to harvest the rye or barley or other grain. You will have to walk on the rice to do this, but don't worry about it. Stack the grain to dry for about a week and then thresh it.

8. Spread the threshed straw and husks on the field. If you have more than one plot, don't spread it in the same plot in which it was grown. If you grew rye in one plot and oats in another, spread the oat straw in the rye field and vice versa.

9. By early summer only rice and weeds are growing. Now is the time to flood the paddy for about a week until the clover turns yellow (without killing it).

10. Over the summer the field should always be at least half rice. Prepare the seeds of the grains for sowing, and pick a different grain to plant in the fall than you did the year before.

When to Harvest

Grain goes through various stages before it is ready to harvest. It starts out as a seed, sprouts, and becomes a seedling with a few short leaves. Then it becomes a *tiller* by sending out a couple of thicker shoots. The stem grows until it begins *booting*, or forming a head, at which point it is *heading*. Heading continues until the head is completely formed, right up until it starts to flower. Flowering completes the pollination process, and that's when the grain enters the *milk* phase. This is when the kernel starts forming and you can squeeze out a milky fluid. Eventually the kernel will dry out and will become mature during the *dough* stage. Ideally grain should be harvested when it is in the

▾ **Ducks in a winter rice field.**

The Ultimate Guide to Permaculture

late *dough* stage, when it is far past the milk stage but still able to be dented. When you let it dry, it will become dead ripe.

Harvesting the Old Way

1. You will need a *scythe*, *sickle*, or horse-drawn *mower*. A horse is the fastest and easiest method, but requires the care and maintenance of a horse. A sickle is for very small areas. Larger areas require a scythe. The scythe is large blade that is swung low to the ground with a long wooden handle called a *snath*. Holding on to the *nibs* (handgrips) of the snath and keeping your arms straight, twist at the waist while stepping forward. That way you move forward and cut at the same time. Keep the top of the snath down so that the tip of your blade won't go into the ground. Try not to cut the weeds and grass. A *cradle* attached the snath will collect what you cut, so it can be dumped into piles. There is a certain art to scything smoothly and efficiently, but with practice the movements become second nature. A *windrow* is a long row of grain; a *pile* is a small heap of grain. The mower will make windrows, while a scythe will make a pile or windrows. Make a pile if the grain is dead ripe, so it can be shocked. If it is not dead ripe, make windrows.

2. Sheave and shock the grain and let dry in the field in the fall. A *sheave* is a large handful of stalks, which is tied near the top with barren tillers (the secondary stalks that grow from a grain plant's base) into a binder's knot. To tie it, twist the stalks together, then tuck under the band going around the sheave. Several sheaves are leaned against each other in the field, and another sheave is set on top of them, making a *shock* big enough for your hands to meet on the other side.

▾ **Milk stage.**

▲ Dead ripe.

▲ Scythe.

3. You may have to harvest early and let the grain dry. You can let it dry in the field as it is, or bring it into the barn, if the barn is large and dry. When there are no green stems at all and the corn stalks sound hollow when you tap on them, it is dry.

4. Oats and barley cannot be hulled by *flailing*, or whacking with a tool called a *flail*. They must be steamed and ground

▲ **Sheaf of wheat.**

▼ **Large shocks.**

in a mill. To *thresh* (remove the hulls of) other grains, use a hand flail to whack the grain while it is spread on the barn floor. Pile the straw separately. A flail has a wood handle attached to another piece of wood that looks similar to the handle, called a *swingle* because it swings around. A loose ring attaches the swingle to the handle. To use a flail, swing the flail back and forth quickly, hitting the grain on the barn floor. When you do all of it, flip the grain all over with pitchforks and flail it again. Do your flailing in the winter. When done, winnow it. For oats use a heavy swingle, for beans a light swingle.

5. To *winnow* (separate the chaff from the grain), toss the grain up with a basket and catch it again outside on a windy day. Alternatively, you can toss it in the air with a pitchfork, or use a winnowing

▲ Flail.

away. The grain must be very dry before packing it and free of any other material except for bay leaves, which can help. Keep the containers in a cool, dry place. Grain will keep for a year or more, until you grind it. It should be checked now and then for mold, bugs, or signs of rodents. Once you grind the grain, it must be used immediately.

9. Grain can sometimes get *ergot fungus*, although it is rare. It happens if your grain gets damp and turns the grain hard, black, and purple on the inside. Don't ever eat moldy grain or feed it to your animals, because you can die and your animals can get sick. If your seed grain gets moldy, you should throw it out, but if you are very desperate, soak the grain in really salty water. The *sclerotia* (ergot masses) will float to the top where they can be skimmed off.

tray, which has a frame with a screen. To winnow beans, wait for a windy day and pour them back and forth between two tubs. Save the chaff for livestock feed, and save the straw or pile it in the field for the animals to eat.

6. Make sure the grain is as dry as possible. Sun dry it if you need to. Then store it.

7. You can save seed from your second crop of grain (unless you're using a hybrid). Your seed should be the best seed heads, unbroken and healthy. Let it dry in the shock for at least a month until it is totally dry. Then thresh it and make sure it doesn't have any leaves or twigs.

8. All grains, whether for seed or eating, must be stored in rodent-proof containers in the house (not the shed), with cats around to keep the mice

Solutions for Bugs in the Grain

Dry ice: You will need 1 tablespoon of dry ice per 5 gallons (19 liters) of grain. Get an airtight container, put the ice on the bottom, pour the grain on top, wait an hour, then seal the container. The grain produces carbon dioxide and kills the bugs.

Heat: You can't use this method for seed grain, but it works for edible grain. Spread a quarter inch (0.6 cm) on a pan and heat it in the oven at 140°F (60°C) for 30 minutes.

Processing grains:

Cracking: breaking the kernel in two or more pieces, usually for corn.

▲ Marble hand grinder.

Crimping: flattening the kernel slightly, usually for oats.

Flaking: treating with heat and/or moisture, then flattening.

Grinding: forcing through rollers and screens.

Rolling: smashing between rollers at different speeds with or without steaming.

Grain can be ground at home with a mortar and pestle, a hand grinder, or an electric mill. The quickest, easiest option is the electric mill because you only have to put it through once. A hand grinder must be cranked for a long time: Each pass through you have to sift the flour and then put it through several more times depending on the texture you want. The coarsest setting on an electric mill makes grits, which can be used to make cereal and animal feed. The finest setting is to make cake flour, which is very fine flour. Generally, 1 cup of grain makes 1.5 cups of flour.

CASH CROPS

The way to become rich is to put all your eggs in one basket and then watch that basket.

~ Andrew Carnegie

Berries

While berries grow in almost every climate, they are most suited to cool and temperate zones, where they grow everywhere naturally. Cranberries, blueberries, raspberries, and strawberries are all pioneers, but unlike most pioneer species, they also offer a valuable edible product. They improve the soil and shelter seedlings from invading deer. They also grow prolifically in soil that isn't very good and need little care except frequent picking. They should be provided with liquid manure and thick mulch to keep the grass from growing. They also need to be protected from birds, which will quickly eat 30% of your cash crop. Berries must be

▲ **Berry cash crop.**

picked every day, and so if you are within reasonable driving distance of a city, you can offer them as a U-pick product.

Bird protection can be done with a mesh cage, which will also house a polyculture system. The raspberries or boysenberries grow on trellises to save space, and blueberries can be intercropped on raised beds 2 feet (0.6 meter) high and 5 feet (1.5 meters) wide with drainage at the base, preferably with piping. Strawberries are grown as a groundcover. Lizards, frogs, and quail are released into the cage to control insects (and the quail can provide food). This type of system must be somewhat small, but because there is very little crop loss, it may be more profitable.

If you decide you want to grow a bigger cash crop and go the U-pick route, bird-deterrent kites are available, which

look like hawks and are simply tethered around the field. Their effectiveness is questionable, but if the kites are removed after the harvest, the birds will hopefully not get used to them. The advantage of having a bigger field is that people will do the work for you, but you must build wider, grassy paths between the 3 feet (1 meter) raised beds. You will also need buckets, scales, and bags for packaging.

Both of these systems use drip lines for irrigation. It is quite common in North America for berries to be watered with sprinklers, which are much easier to install, and possibly cheaper, but during the summer you will be watering twice a day. Sprinklers must be left on for hours, letting much of the water evaporate in the sun, and some of it won't even make its way down to the soil. Drip line uses less water and

conserves it by depositing it directly near the roots.

Blackberries are tasty but also very likely to get out of control. It is probable that there are some blackberries on your property already, which you may want to utilize as a hedge or barrier somewhere, but you will have to form a strategy in keeping them under control. Removing them takes years, so it's a good idea to start right away. In a very small area it is possible to cut them back and cover them with a strong mulch of tough plastic weighed down with rocks or other material. Once they have rotted there for 2 years, you can dig the roots up. An area of blackberries a quarter acre or more can be fenced off and used as a pig field at a ratio of 20 pigs per acre (0.4 hectare). The next year 12 goats per acre (0.4 hectare)

are grazed there, followed by pigs again the next. The blackberries won't return as long as you keep something in there, like sheep or goats, or plant trees or a hay cash crop. Another strategy is to plant apple, fig, pear, and plum trees in the middle of the blackberries 40 feet (12 meters) from the edge. In about 5 years these trees will have grown enough that you can let cattle in to graze. The cattle will eat the windfall fruit and trample down all the blackberries.

Hay

Alfalfa makes one of the best choices for hay, especially when mixed with bromegrass. It produces at least 5 tons (4.5 metric tons) of hay per acre (0.4 hectare) for 3 cuttings a year, and is susceptible to alfalfa weevil that makes it a bit trickier to

grow but is still worth it. This is the most popular hay and in rural areas is in demand by horse and livestock owners.

Red clover and timothy mixed hay is the second best kind. They produce 2.5 tons (2.2 metric tons) per yearly harvest. It grows slower so you can't do three cuttings, but it doesn't get alfalfa weevil.

Millet grows very quickly, can be fed 30 days after planting, and gets 5 tons (4.5 metric tons) per acre. It must be fed with buckwheat for full nutrition, so it's a good idea to sell it together. Millet has recently become more popular as a grain for making bread.

Oat straw makes the most nutritious hay of all and likes cooler weather.

To plant hay, you can *broadcast* (toss the seeds evenly by hand) on snow in February or on frozen ground in March. Cut the clovers with a scythe or mower just as

they begin to blossom. You have to let the hay dry in the field before putting it in the barn. Rain destroys hay. It will probably take 2 days after cutting for it to dry, unless the weather is very hot and windy. When the hay is almost dry, rake it into windrows with a hand rake or a horse-drawn hay rake. To test if it's dry enough, take a 2-inch (5-cm) thick bunch and twist it. It should break on the third twist. Use a pitchfork to load into a hay wagon, and take it to the barn.

Eggs

The general consensus among free-range chicken owners is that you need at least 100 chickens to make any kind of money selling eggs. A hundred chickens that lay a decent number of eggs, up to 300 a year, will give you 3,000 eggs. These are usually sold for a couple of dollars (US) per dozen in rural area and slightly more

in a suburban location. If you take away your own share, you could make a couple of thousand dollars per year on them. This kind of profit is only possible if they are free range; otherwise, the revenues will be eaten up in feed costs.

Check your local laws regarding selling eggs. Usually, law requires that the eggs be clean, and if you are selling over a certain amount, then they must be *graded*, or given a letter grade based on their quality. They don't usually have to be actually washed or put in a new container, but they do have to be refrigerated.

Selling eggs in a free-range setting isn't an endeavor that will pay your mortgage, but it will provide some extra income, especially since your chickens will be producing more eggs than you can use, and you have the infrastructure anyway.

ZONE 4: PASTURE MANAGEMENT

The investigation of nature is an infinite pasture-ground where all may graze, and where the more bite, the longer the grass grows, the sweeter is its flavor, and the more it nourishes.

~ Aldous Huxley

Forage System for Cattle and Sheep

In most temperate climates, 20 acres (8 hectares) is usually enough land to raise enough livestock to produce a small income. The quality of your pasture is the key to this. The pasture should be able to feed your animals through a drought, protect them from storms and sun, restore the soil, and prevent erosion. The pasture provides grasses and legumes, seedpods, sprouted grain, silage, and tree leaves. However, it is a tricky process to feed animals the entire year in this way. The herd must be *culled*, or thinned out by selling or butchering the young males, but even then there will still be a food shortage in the winter. To cover that shortage, tree crops are used.

It takes at least 5 to 10 years to switch to a forage system with the right proportion of trees, but the benefits are enormous. Not only does it save tremendous amounts of energy on the part of the farmer, but also the cattle are healthier and happier, which means higher yields. The first year, about 10% of the land should be trees and bushes, planted in the same community planting method used with all the other gardens in the other zones. By the fifth year, the proportion should be 40%, and sheep can be allowed to forage, perhaps along with a few young cows. They can be allowed to browse for a short while and then removed.

As the years go on, you can allow them longer and longer periods of grazing time.

Forage trees include fig, poplar, willow, chestnut, oak, and pine. Bamboo can be either a forage or a timber crop.

Large Grazing Pasture

Even on a property with a very large grazing area, the fenced perimeter can provide protection and forage. A second fence should be built inside the first, and the space between should be planted with trees, hedges, and spiny shrubs. This will provide protection, windbreak, fruit, nuts, wood, and forage for sheep as well as bees and birds.

An area of 50 acres or more can also still be developed into a forage system:

1. Fence off a small area you want to develop. You will probably have to use an electric fence.
2. Go through the process of rehabilitating the soil with a chisel plow

and lime, as described earlier in this book.

3. Plant a small group of windbreak and forage trees in the very center. Mulch and fertilize them well. Initially, small seedlings may need to be protected with shelter, such as a tire with mulch in it.

4. Introduce ducks and geese into the area and make sure they don't do any damage.

5. After time, when the area has established itself, move the fences to the area next to it and go through the same process.

6. Go back and cut out trees that are doing badly, leaving only the strongest and highest yielding trees and shrubs.

Can I fatten them up?

Traditional farming has typically used the strategy of cramming animals with concentrated foods (like grain) to make

Timber in the Pasture

To save space and create shelter and windbreaks, not only are trees planted for the animals to eat from, but for firewood and building materials as well. These should be planted to fit the contours of the land and allowed to mature enough that animals won't damage them, which can take 30 years. Animals can be allowed in to graze before the grass is harvested for hay or whatever cover crop you are growing there.

them gain weight as quickly as possible. You will have to use concentrated foods in a permaculture system, but not for weight gain. They are for fattening up animals just before butchering, for maintaining milk and egg production, and for keeping animals going during times when forage is lacking.

The feed shouldn't be bought, however—it should come from the land. This is so everything from the soil can eventually cycle back to where it came from.

These feeds include acorns, chestnuts, wheat, buckwheat, oats, barley, peas, chickpeas, pumpkins, sunflower seeds, and rye. Most of these can and should be sprouted before being fed to animals, the same way they should be for people.

WOODLOT

He lowered the window, and looked out at the rising sun. There was a ridge of ploughed land, with a plough upon it where it had been left last night when the horses were unyoked; beyond, a quiet coppice-wood, in which many leaves of burning red and golden yellow still remained upon the trees. Though the earth was cold and wet, the sky was clear, and the sun rose bright, placid, and beautiful.

~ Charles Dickens

Coppicing

Fallen branches, thinned trees, and old wood will provide the bulk of your fuel when you first start using your woodlot for firewood. This supply would soon be exhausted, however, if you do not plant new trees frequently, since you will be cutting down around a quarter of the trees every year, and none of them will last longer than 7 years. Choose species based on their ability to grow quickly. This type of woodland is known as a *coppice*, an area that is periodically cut back to ground level to stimulate growth and provide firewood. These species are also chosen for their ability to regrow from a stump. The tree is cut down close to the ground and sprouts up again quickly the next year out of the stump, providing new wood in a relatively

The Ultimate Guide to Permaculture

short amount of time (7 years instead of 40 or more).

Coppicing promotes the diversity of species because it lets in light to the forest, creates microclimates, and surprisingly makes the trees live longer. Species used for this purpose include maple, pawpaw, birch, beech, ash, oak, alder, willow, hawthorn, persimmon, ginkgo, honey locust, and black locust.

Coppicing isn't especially difficult to do, and in fact because the tree is smaller than a large piece of timber, it is easier to cut down. The tree must be cut as close as possible to the part where the trunk begins to widen out to the roots. Each of the tree sections (usually there's not just one trunk) must be cut at an angle of 30° facing out of the trunk so that the rainwater runs off onto the ground.

You will want to plant a mixture of polewood, high-quality timber, and hedges. The polewood will be used for fencing, outbuildings, and carpentry, and the timber can provide food and possible cash crops, such as walnut or bamboo. The hedges and shrubs keep grass at bay and establish a microclimate for trees to flourish.

Although the timber trees won't necessarily be coppiced, you can use the forest space to grow sugars for food and alcohol production (and possibly for biodiesel as well). Species of trees that provide fruit or sugary sap are also planted amongst your other trees. Sugar cane or beets can also serve this purpose as well, but trees are particularly valuable for the little care that they need. This sap or fruit juice could also be distilled into alcohol. Any waste product from this process should be returned to the forest as mulch.

The trees are cut when you need them, at a minimum of 7 years and usually no greater than 25 years old. The longer you wait, the larger the timber.

ZONE 5: THE WILDS

In wilderness is the preservation of the world.

~ Henry David Thoreau

When land is so precious and food production so vital, it may seem nonsensical to leave an area of your property untended and allowed to fall into so-called disarray. This natural state is exactly what is needed for wild creatures to live comfortably, and it is the best possible solution to erosion. If you are on a slope and have situated your zones properly, Zone 5 will be likely near the top of the slope, where soil is at the greatest risk of erosion. Wild forest is the source of most of your air, clean water, medicinal breakthroughs, building materials, and diversity.

Zone 5 is also the location of your wildlife corridor, and it might cross some other zones of your property if you find that deer are constantly walking across a certain spot. Rather than always being at odds with animals eating your gardens, fence off their path separate from the rest and leave it wild with natural forage for them to eat, like an animal highway through your property.

There's really not much to say about Zone 5 other than it will be very tempting to mess with it. It must be left alone.

▾ This area is left relatively undisturbed.

8 | Community

I am of the opinion that my life belongs to the whole
community and as long as I live, it is my privilege to do for
it whatever I can. I want to be thoroughly used up when I
die, for the harder I work the more I live.

~ *George Bernard Shaw*

LIKE-MINDED PEOPLE

The antidote for fifty enemies is one friend.

~ Aristotle

What Community Is (and Isn't)

Historically, permaculture has been heavily focused on the building of *community*. The phrase "community" in the permaculture sense is not the type of neighborhood that springs up on its own, a group of standardized houses built in subdivisions with cul-de-sacs to play basketball in. You had friends and acquaintances in your neighborhood, but very few of us experienced that *sense* of community that so many people crave.

Community, as it should be, is intentional. It is a more connected way to live, encouraging greater involvement with people outside of your immediate family. It is also intended as a support system that takes care of basic human needs outside of what a family can provide—business and income, emotional needs, services and products, child care, insurance against disaster, etc. This type of community requires better communication and more time investment than the fenced-off chitchat neighborhoods we have today.

Permaculture encourages experiments in alternative social constructs, political policy, banking, as well as religion and relationships, but these experiments have been less successful than the establishment of a *village*. There are now many well-established small communities, or eco-villages, that have been built on the ethics of permaculture, with varying degrees of communal sharing.

We are not talking about communes, however. A commune is an organization of people who share all of their income and sometimes their homes. Every commune ever established has either failed or changed its structure, which does not necessarily mean that communalism is wrong or somehow doomed, but it does mean that it would take a special group of humans to pull it off. The villages described by permaculture are more akin to a regular town but described with utopian ideal. A permaculture village has all of its energy and resources provided locally through agrarian projects, and all of the people are able to work at employment that is meaningful, localized, and ecological. All ages are provided for adequately, and social services are run locally and independently. At its heart, permaculture tends to lean towards anarchism in its purest definition: the abolition of a centralized government and the organization of society on a voluntary, cooperative basis without recourse to force or compulsion.

It is for this reason that many scoff at permaculture as optimistic, idealistic, and extreme. These doubters are probably right. In my opinion, humans cannot, and will not, ever create thousands of totally autonomous communities of people that are independent from any outside resources or influences. That's not very optimistic of me, but it is practical. Jobs, health, income, and technology like Internet and computers, these all at some point have to come from somewhere else. A certain percentage of people in a village will be injured or get sick. Modern medicine may have its problems, but we can't ignore the fact that it is responsible for the long lives we enjoy today. These injured or sick people must see a doctor at some point for something, and that doctor has to be trained by a school in another place. Whether the student's tuition was paid for by his own labor or through social programs, a village doesn't have the kind of resources to train a doctor for ten years and is unlikely to possess the knowledge of treating serious injuries and diseases. When that person becomes a doctor, can the village even build a hospital with all of the state-of-the-art equipment needed?

It is interesting and fun to experiment with our culture and social constructs, but in order for permaculture to succeed as a solution to the various ills of society, it must let go of political and religious ideology as part of its overall goals. Permaculture is about designing a healthy, sustainable manner of living for a family or a very small community, and that's it. When the lifestyle of enough individuals becomes conscious and deliberate, it is not necessary to branch out into the bigger picture. The world will inevitably change on its own.

And that brings us back to community. Today, it is very difficult to purchase land, or to use land in an urban environment for permaculture activities, without great expense or annoyed neighbors. To this end, a community must be built up of like-minded people who can pool their resources to acquire land or create protection around the activities of the individual who practices them in an urban environment. This community will never become autonomous (and in fact is most likely to be in the middle of a city and completely dependent on that urban environment), but its level of self-reliance will be much higher than it was before. The community will also be able to more effectively follow the third permaculture ethic: Share the surplus. Individuals will share with their small community of friends, who will share with the greater community of not-so-like-minded neighbors. That is when real social change begins.

To summarize, permaculture applied to community has three goals:
- Create a support and teaching network for like-minded people to be able to change their lifestyle.
- People do not have to live together or have anything in common other than the belief that humanity should live sustainably and self-reliantly.
- Self-reliance does not come from independence from people, but rather from independence from business and organizations that keep you alive by bringing products and resources from far away places.

Starting a Community

It is recommended in the official permaculture certificate course curriculum to avoid business meetings, and that consensus in every day decisions is unnecessary. I disagree with these strategies because they increase the risk of collapse in the community. Frequent communication is important even in the

▲ **Campfire meals build community.**

smallest group. Everyone needs to be on the same page, and any decisions that impact everyone else should be made in consensus. This is a hassle and is incredibly challenging, but it is also necessary. There always comes a point during a community's evolution where one person feels left out of the process. As soon as that individual feels excluded (whether the exclusion is real or imaginary), this perception will stir up turmoil. Turmoil impedes progress, and the group must be moving steadily forward towards a preplanned goal in order to continue as a community. Otherwise, people naturally lose motivation or run out of energy. One discontented person can cause the collapse of the entire process.

One effective method of making communication more efficient is ensuring that groups stay small. That first core group of people was brought together by a common purpose, but as soon as more people join, things get complicated. All of those extra people have their own reasons for being there. These people should be formed into small groups according to their individual personalities and interests, so that each has something smaller to manage. Each group should have a spokesperson and a secretary. The secretary keeps meticulous records of everything said and done, and the spokesperson reports to the greater community. This may seem very formal and traditional, but this "committee" structure has proven itself to be functional even when individuals aren't very experienced with it or have personalities that are difficult to work with.

Not everyone gets along, and no one communicates effectively all of the time. Creating an environment that is honest about this basic fact and works around it is the key to success. Flexibility is part of this. When a job is done, people can move on to something else. When two people aren't working well together, rather than taking it personally, they should recognize that sometimes personalities clash, and they need to move on. One of the greatest tools for helping people learn to do this is through a common mission. A mission is more than a goal—it embodies the values and direction of the group. The mission must be decided as a community and written down, and everyone should be reminded of it frequently. Every decision made by the community should be held up to the mission statement to make sure the group is sticking to its values.

Dissent is expected. Group consensus is difficult because if someone disagrees with the group, the whole group cannot move forward. However, there are very few decisions in permaculture that must be made immediately. A single individual may slow down a decision for quite a while, but that is one of the drawbacks of community that must simply be tolerated. It is very important that the founders of the community have communication and consensus decision-making training so that they can facilitate the communication of the group. There are many invaluable courses available that are specifically tailored to consensus organizing for intentional communities.

The other greatest tool in community is food. Just because you are getting together for a business meeting to discuss difficult topics doesn't mean you can't do it over an amazing potluck dinner of homegrown food.

VILLAGE DESIGN

In the end, poverty, putridity and pestilence; work, wealth and worry; health, happiness and hell, all simmer down into village problems.

~ Martin H. Fischer

The Decision to Live Together

Today the most common reason for people to decide to live together in North America is for land access. Urban agriculture is on the rise, but there are limits to what you can do. Some of those limits are imposed by the city to keep the town clean, and people push those boundaries, but simply having rural land is much easier. Land is expensive, and also a lot of work, and so sharing it makes sense.

Once a group of people decide to buy a piece of land together, they must then hire a lawyer. The land is divided into areas for living, agriculture, business, and wilderness, and some of these areas are put into a legal *trust*. A trust is a way of preserving land without an individual having to be alive to own it, and the trust must be registered like a non-profit organization. The areas for living are often rented out to the people of the group or bought as shares, like a corporation.

The decision to live together shouldn't be taken lightly. Once you invest in a trust, there is usually no way to recoup your money should you decide you don't like these people after all. Some very lucky people have been able to convince their city to provide acreage for farming as a benefit to the community. In comparison to the organizational and legal struggles of forming a trust, a city land gift is incredibly less difficult to envision.

Placement of Houses

Groups of houses for a community should be placed in the same way that any house would be in a permaculture design. Consideration should be made for the wind and sun, the temperature, etc. However, because you are dealing with more than one house, space becomes an issue.

The Ultimate Guide to Permaculture

1. Houses in a community should be placed in the shelter of warmer sides of hills and follow the contour of the land.
2. There should also be a balance between privacy and community. This is possible by making small clusters of houses grouped around a common area. Usually, about 5 to 8 houses per cluster works well for people.
3. Every house should face the sun and be built according to passive solar principles.
4. All water runoff from the houses should lead to swales planted with trees and shrubs.
5. Rather than large front yards divided by fences, clusters of houses should have tiny front yards with tall hedges for privacy facing a narrow street. The space saved by crowding these aspects together can be put towards the common area, which can be a small park, an orchard, or a garden.

▾ Central community garden.

BUSINESS

To practice Right Livelihood (samyag ajiva), you have to find a way to earn your living without transgressing your ideals of love and compassion.

~ Thich Nhat Hanh

Right Livelihood

Buddhism has coined a term that accurately sums up how to apply permaculture ethics and principles to one's means of employment. *Right livelihood* is part of the principle teachings of Buddha, as one of the key components of the Noble Eightfold Path. It means that people should refrain from occupations that either directly or indirectly result in harm to another living being. In permaculture, this is taken one step further to include jobs which have no meaning or take unnecessary time away from one's family. It is true that not everyone

can leave the job he or she is unhappy with (and it is probably true that not many people *like* to work at all), but with effort, it is possible for creative and hardworking people to achieve a career goal that could be described as right livelihood.

First we have to redefine career, however. People must let go of the idea of having a "career" for the sake of having a career. Anything that contributes to society and the welfare of the community should be considered a successful career, whether someone is a doctor or the person who dumps the sawdust in the compost toilet. Today, especially since a real-life community is so difficult to establish, we have the Internet. On the Internet, the whole world is at your fingertips, there are no real rules, and you are free to capitalize on knowledge and services that you have. I believe that using all of the tools available on the Internet is the first step towards an independent career.

Many people reading this book may be interested in utilizing permaculture for profit, through growing food, or offering

education, especially if they are trying to form a community. Once again, the Internet is the place to start. See what other people are doing, then offer something unique that they are not.

Cash Crops

One person alone can't maintain a true productive farm that is growing a serious cash crop, which is why this section is in the Community chapter. An entire family has to be involved, or at least a group of seasonal helpers. In choosing a cash crop, pick one that is low in bulk, such as honey or berries. It should also be easy to process without much equipment, or be *value added*. Value added products are jam or butter—raw materials like berries and milk that are processed into something people want, which adds value. It's a good idea to have a second cash crop that is something non-perishable, like firewood or nuts, which can be sold throughout the year. Some cash crop ideas are:

▲ Herbs are a very profitable agricultural business.

Aquatic nursery: Fish, bee and duck forage, friendly insect plants, or ornamentals.

Berries: Fruit, u-pick service, or plant nursery.

Rare plants: Useful permaculture plants or bee, bird, and beneficial insect forage.

Seeds: Rare or unusual heirloom seeds.

Animals: Geese, silkworms, earthworms, bantams, milk or mowing goats, draft horses, heritage cows, specialty sheep, or quail.

Hedges and trees: Specialize in local species, trees for regenerating forests, windbreaks, animal forage, bamboo, or food crops.

Organic food: Typical fruits, vegetables, nuts, milk, eggs, wool, meat, or flowers.

Value-added food: Smoked meat, dried fruit, jam, feathers, dried flowers or wreaths, or pickles.

Craft supplies: Willow, bamboo, natural dyes, or wool.

Natural pest control: Prepared powders or sprays, or nursery plants like marigold or yarrow.

Herbs: Natural medicinal and beauty preparations, teas, or dried herbs.

Tourism: Farm holidays, camps, retreats, workshops, or classes.

Marketing

Organic farms today sell most of their products in three ways: at farmer's markets, direct to consumers through u-pick or community supported agriculture (CSA), or direct to restaurants. The CSA is by far the most rapidly growing method of direct sales, and its the principles are simple. Each family purchases a *share* every year, usually priced somewhere around $300–$500 for the season, and in return receives a weekly box of fresh-picked vegetables and fruits, and often some value-added products. These are delivered to their door or to a set delivery spot for pick-up.

The struggle with starting a CSA is managing quantity and quality of the product. Some CSAs start out the first year hoping that the weather will be good and they can fulfill their promises, only to produce much less food than they expected. They end up providing very little food weekly, and customers feel as though their up-front investment was not worth it. A startup CSA can't get greedy by overselling shares to too many people. Start out by providing food to your family, and then expand to a couple more families. Make sure the weekly food box you provide is overflowing with food, with extra surprises thrown in for good measure. These should be high quality and fresh as well.

The same strategy holds true for any business. Start small and make sure you can provide a good product so that customers will recommend you. If a customer isn't happy, do anything it takes to make him or her happy.

Commercial Greenhouses

While you already have a greenhouse by your house for your own food, a low-effort but somewhat more expensive cash generator is a commercial greenhouse. It is much cheaper to build one, but take into consideration the work and potential risks if you do. It is common for businesses to start out by building a *high tunnel*, which is simply a long framework, made of PVC piping, that holds up a light plastic cover. This plastic cover must be removed in the winter, so this is not for year-round growth. This type of

▾ **Commercial high tunnel.**

greenhouse is cheap, but it is also very susceptible to wind damage.

It is much easier to buy a real commercial greenhouse than to build one which will have built-in features like ventilation and be sturdy enough to function all year round. One feature it needs to have in a cool climate is a separate entry room. Every time the door opens, heat is lost, so an entryway with two doors is crucial. The first crops you grow should be designed to pay for the greenhouse, which means growing high value products like tomatoes, peppers, and fruit.

Of course, while running a commercial greenhouse business is most often done with a single high-value crop, in a permaculture system you would still follow the principles of community growing and polyculture. Pick at least three crops that can work together, and consider using aquaculture tanks or birds as well. This kind of business is more labor intensive than your ultimate goal with your personal food production, but that is the nature of large-scale agriculture.

COMMUNITY RESOURCES

Where is the wisdom we have lost in knowledge? Where is the knowledge we have lost in information?

~ T. S. Eliot

Food Production

At a community level, food production should be divided into two categories: family and community. One of the reasons why communal systems fail is lack of ownership. When individuals don't feel any personal investment in something, they don't work efficiently or well. Each family should be mostly responsible for its own food production, with room for its own garden and animals. However, this will only provide food enough for each individual family for the summer. It will not provide enough to make a profit, and probably not enough to get everyone through the winter.

▼ Urban agriculture is a necessary part of future city planning.

The community garden and orchard will provide winter storage and cash crops. Animals can be kept together, even if each family has its own animals. It is common to require each family to contribute a certain number of hours to the community in return for the privilege of living there.

Recycling

The management of waste is a tremendous task even in a small community. It is a massive waste of energy to sort all of the different materials generated by all the households of a community in one place, so all waste material should be sorted at the source. Each family should be responsible for sorting its own waste (glass, paper, metal, plastic, etc.). Each family should also have its own composting bins for organic materials from the kitchen and small yard waste. The community should also have a facility for large materials like tree branches where these can be chopped up and composted. Once all of the waste has been sorted and collected, it can be sold to manufacturers for recycling or repurposed for other uses.

Access to Land

Each community should promote access to agriculture, no matter how urban the community may be. This includes community gardens, farm co-ops and CSA membership, garden clubs, and urban farming.

Land access is one of the most worrisome situations facing North America today. There are acres and acres of empty land, but very little of it is currently *arable*. Arable land can be used for agriculture because it has been cleared of trees, the soil is fertile, and it is relatively free of

▲ Precious agricultural land is subdivided and sold.

rocks. The current amount of available arable land is rapidly decreasing due to urban sprawl and environmental damage. Some land can be made arable, but the process is difficult and takes a long time. This is why farmland prices continue to rise to astronomical levels that, ironically, only developers can afford. This problem is made worse by the fact that half of the farmers in America are retiring very soon and are not passing their land on to their children. Without a retirement fund, they have to sell off the land to pay for the right to rest in their old age. This vicious cycle

The Ultimate Guide to Permaculture

▲ Big suburban houses encroach on arable soil.

has resulted in the outsourcing of many food staples to other countries like China and Mexico, leaving North America's food security in great jeopardy. Every person with arable land, no matter how small, should be growing food or sharing that space with someone who doesn't have the same resources.

9 | Plants

What is a weed? A plant whose virtues have not yet been discovered.

~ *Ralph Waldo Emerson*

Note: The following tables and guides are by no means complete. A variety of different climates and conditions are represented, and they should give you a starting list of species with which to experiment Contact your local agricultural extension to find out your best native plants, and use these lists as a guide to what to look for. Also, just because a plant is listed as edible doesn't mean every species can be eaten or that the whole plant is edible. Some plant groups may only have one edible species, and some are only palatable to animals. Read the notes for each plant and make sure you have the right species.

ALPHABETICAL PLANT GUIDE

Acacia

Uses: bare soil erosion control, edible pods, wood and timber, dryland tolerant
Species: Prairie Acacia (*Acacia angustissima*)
Growing: Be aware that there are similar-looking acacia relatives which are poisonous; be sure about your plants. Prairie Acacia can provide animal fodder and takes the abuse of frequent cutting, but it can't become a primary source of food. Too much of it can be toxic to some animals. It is not generally edible to humans. It is a perennial that does well in the southern United States, where the weather is warm and dry. It is tough and hardy and doesn't mind hard, dry, or rocky soils.

Alder

Uses: leguminous nitrogen fixer, wood
Species: Italian Alder (*Alnus cordata*), Speckled Alder (*Alnus rugosa*), Grey Alder (*Alnus incana*), Smooth Alder (*Alnus serrulata*)
Growing: Alders grow natively in most areas of North America because they come in such a variety of species. The Gray and Italian Alders are dryland tolerant, while Speckled and Smooth require wet conditions. They enjoy partial shade to full sun. The Italian and Gray species get quite a bit taller, up to 60 feet (18 meters), while the Speckled and Smooth stay around 20–35 feet (6–11 meters) tall. Dryland tolerant alders can be used as part of a forest garden with fruit trees if they are thinned out regularly.

Alfalfa

Uses: bee forage, edible, leguminous nitrogen fixer
Species: Alfalfa (*Medicago sativa*)
Growing: Alfalfa is the forage and hay of choice for livestock, although too much of it can be fatal if an animal is suddenly switched to it exclusively. People

eat alfalfa sprouts too, which are incredibly healthy. Alfalfa is a perennial legume that grows well in well-drained loam. Wait until the alfalfa just starts to bloom before you allow the animas in to forage, or you cut it for hay, and don't let it get shorter than 2 inches (5 cm). Let it recover for another 6 weeks, then do it again. Make sure you cut it a month before the first frost.

Almond

Uses: nut, bee forage, dryland tolerant

Species: Sweet Almond (*Prunus dulcis*)

Growing: There are many varieties of almond, and most of the North American ones are grown in California, where they enjoy the warm sunny climate. The Sweets are the edible kind, but it's just not worth it to grow the Bitter variety, which you can't eat. Some varieties are not self-pollinating, so you would need a few trees. It generally takes 5 years to see any kind of real production. They only grow around 15–30 feet (5–9 meters) tall. They like plenty of water since they are usually grown in dry and hot areas, and they may be susceptible to peach leaf curl, a fungus that turns the leaves brown and curls them. Remove the infected leaves and burn them. The tree will produce small green hulls that will begin to dry and drop off the tree, and you will probably have to pick some of them. Remove the almonds from the husk and let them dry. When they are fully dry, which takes a few days, the nuts will rattle inside the shells. They can be eaten raw or roasted.

Amaranth

Uses: edible leaves and seeds, dryland tolerant

Species: Purple Amaranth (*Amaranthus blitum*), Red Amaranth (*Amaranthus cruentus*),

Growing: Amaranth is treated like a grain, although it's not a grass. It has broad leaves, which are some of the healthiest leaves you can eat, and is full of protein. Amaranth is dryland tolerant and enjoys well-drained soil and full sun. It grows best in the southern United States. For a similar plant that can be grown in northern areas, choose Quinoa. To harvest the seeds for grain, you must rub the flower heads to see if the seeds fall out easily. Sometimes the plants continue flowering long after the seeds are formed, so they aren't a very good indication of readiness. Harvest the seeds by shaking and rubbing them into a bucket. You will need to thresh away the hulls like any other grain and blow away the chaff. The simplest way to do this is by rubbing the flower heads on a large screen and then using a fan to blow the chaff away. Use the same screen trays to dry the seeds, either in the sun or near the wood stove. Store in a cool dry place. Amaranth can be cooked just like rice or ground into flour and added to bread.

American Linden

Uses: bee forage, wood and timber, edible

Species: American Linden (*Tilia americana*)

Growing: Traditionally, the American Linden (or Basswood) was used to make fiber for baskets and nets. The leaf and bark can also be used in traditional medicine. It can be grown in many climates and temperature ranges and enjoys rich, well-drained soil. It serves as a valuable source of timber, as long as the seedlings are protected from deer and other nibblers.

American Yellowwood

Uses: leguminous nitrogen fixer

Species: American Yellowwood (*Cladrastis kentukea*)

Growing: Native to most of eastern North America, the American (or Kentucky) Yellowwood is a tree growing 30–50 feet tall that enjoys full sun. It is fairly hardy and doesn't have many pests. It needs to be pruned when it is young to prevent some of the larger branches from growing larger than half the diameter of the trunk, and the branches should not be clumped together. It doesn't have any enemies.

Apple

Uses: bee forage, edible fruit, wood and timber

Species: Apple (*Malus domestica*)

Growing: There are over 7,000 kinds of apples that grow in every climate. Some are for eating, some are for cooking, and some are for drinking. They must be grown in groups because they are not self-pollinating, and they are often grafted for this reason. They are susceptible to pests and diseases, and without pruning they grow to amazing and un-pickable sizes. To counteract this, the home grower can pick a dwarf variety, which is much easier to manage.

Apricot

Uses: bee forage, edible fruit

Species: Apricot (*Prunus armeniaca*)

Growing: Apricots are hardier to cold temperatures than peaches, which means they can grow in cold climates, but they are also susceptible to spring frosts which can kill the early flowers. They are sensitive to soil type and need fertile, well-drained soil. You will need several trees, as they are cross-pollinated. They won't bear fruit for 2 to 3 years, but you can increase production by pruning after the second year to remove any crossed or rubbing branches. Keep the pruning simple and to a minimum, and do it on a hot day in the summer to prevent sickness. They are not drought-tolerant and may sometimes need some watering, but their roots should never sit in water for too long.

Asparagus

Uses: stream bank erosion control, edible roots/shoots

Species: Asparagus (*Asparagus officinalis*)

Growing: A perennial that enjoys full sun, asparagus is delicious but finicky and not able to compete with other plants. It needs to grow alone, with the exception of tomatoes, which repel asparagus beetles. Some people are allergic to the raw shoots, but they are better cooked anyway. Asparagus is a high-nutrition food and worth going to a little trouble for. It is cold hardy and prefers well-drained soil. Traditionally, asparagus is grown in a trench about 1 foot (30 cm) deep, with a bed of 3 inches (8 cm) of composted manure or mushrooms at the bottom. Adding manure tea after it is planted is also beneficial. It needs regular watering and weeding, but with proper care will keep growing for 15 years. You can begin harvesting asparagus spears the year after planting, in mid-spring when the weather gets a bit warmer and drier. Pick only spears that are

7–9 inches (18–23 cm) tall with closed tips. You can only pick them for 4 weeks, and you must make sure you pick them before the tips fan out, as this makes them susceptible to those pesky beetles. When picked, throw them in ice water to chill them, and then put them in a plastic bag in the fridge, where they will last a couple of weeks. They have deep and quick-growing roots that can be used to stabilize stream bank soil.

Autumn Olive

Uses: bare soil erosion control, edible fruit, bee attraction, windbreak, leguminous nitrogen fixer

Species: Autumn Olive (*Elaegnus umbellata*)

Growing: This is a hardy small tree that produces a small, nutritious fruit that can be eaten fresh or dried or turned into jam. The tree also improves the soil by fixing nitrogen, and since it grows in most soil conditions, it can stabilize eroding bare soil and act as a windbreak. Despite its value and versatility, in North America it is often considered an invasive plant or weed because it is difficult to remove. Once established, it will continue to grow back from the roots and cause trouble for other species. Because of this, it should only be grown in places where it already grows, which is chiefly in Eastern North America.

Avocado

Uses: edible fruit

Species: Avocado (*Persea americana*)

Growing: The avocado grows 70 feet tall in a frost-free climate. It is not tolerant to wind and needs deep, well-drained soil. It is also susceptible to diseases. The fruit must ripen after it has been picked. Typically, the fruit simply begins to fall off the tree and would ripen on the ground if you don't gather it up. Pick and store the fruit at room temperature until it is ripe, which usually only takes a few days. The fruit is eaten raw and in a variety of dishes, including desserts. Avocados have high amounts of healthy fat and are very nutritious. Be aware that the plant and pit are toxic to animals.

Azolla

(see Duckweed)

Bamboo

Uses: bare soil erosion control, streambank erosion control, edible roots and shoots, wood and timber

Edible Species: Sweetshoot Bamboo (*Phyllostachys dulcis*), Stone Bamboo (*Phyllostachys nuda*), Blue-green Claucous Bamboo (*Phyllostachys viridiglaucescens*), Temple Bamboo (*Semiarundinaria fastuosa*)

Growing: Bamboo enjoys partial shade to full sun and can be used for almost anything. It can quickly take over a garden and thus needs to be contained, but it is very hardy and does not need any work. The shoots pop out of the ground on their own, but they need to be cooked.

Choose shoots that are heavy and firm, with a thick outer skin. Peel that skin off and cut off the root end. Put it in a pot of water and bring to a boil, then turn it down to a simmer for about an hour or until it is soft through the middle. Turn off the heat and let it just sit there until it is cool. There is probably some extra skin still on that needs to be peeled off, but it should be easy now. Keep the cooking water and use it to store the bamboo in, so it won't dry out. The shoots can be eaten cold on rice or used in stir-fry.

Beans

Uses: edible, leguminous nitrogen fixer, climbing vine

Species: Common Bean (*Phaseolus vulgaris*), Scarlet Runner Bean (*Phaseolus coccineus*), Winged Bean (*Psophocarpus tetragonolobus*)

Growing: There are three kinds of beans: snap, shell, and dry. The snap varieties can be harvested when the pod is developed but the beans are small, and they are eaten whole. The shell varieties must be picked when the beans are more developed but not dried. Dry varieties are picked when the beans rattle in the shell. They can grow in most soil types and need something to climb on. Dry beans need about 3 to 4 months to reach an acceptable level of dryness, which usually happens by early September. Even then, let them sit on a drying rack in a warm, dry place for a couple of days before shelling. Remove the shells, blow the shells (chaff) away, and pick through and remove any other debris or bad beans. Bad beans are moldy or discolored, or have holes in them. They should be so dry that you can't dent them with a fingernail. Other kinds of beans can be harvested continuously throughout the season, and in fact the more you pick the more will grow. The Scarlet Runner Bean is unique in that it grows a big tuberous root and tolerates colder temperatures,

but the growing is pretty much the same. Keep picking it as it grows. The Winged Bean is a quick-growing vine that needs to be given lots of space to climb and a strong trellis. It prefers hot, humid places, but there are varieties that can grow just about anywhere, and the entire plant is edible. The pods are picked continuously, the young leaves can be used like spinach, and the root can be cooked like a potato. The root is full of protein. The beans can also be dried.

Bearberry

Uses: edible fruit, ground cover

Species: Red Bearberry (*Arctostaphylos rubra*), Alpine Bearberry (*Arctostaphlos alpine*)

Growing: Bearberry is a small evergreen shrub that is used as a ground cover. It has the additional benefit of producing edible berries. It enjoys full sun, tolerates most soil types, and is drought-resistant. It does prefer a cooler climate and is beautiful for adding greenery in the winter. The first few years it will grow very slowly. The

▼ **Bearberry**

berries aren't particularly tasty, but traditionally they were used for herbal medicine, dried and ground for mixing into other foods, or turned into jam.

Bee Balm

Uses: bee forage, edible

Species: Bee Balm (*Monarda fistulosa*), Scarlet Bee Balm (*Monarda didyma*)

Growing: Bee Balm has many traditional medicinal uses and is easy to grow. It thrives in most areas and simply needs regular watering. Cut it back once a year to keep it between 4 to 8 inches (10–20 cm) tall. It can become an aggressive spreader, and so dividing the plant every 2 to 3 years will keep it both healthy and in check.

Beet

Uses: edible roots

Species: Beetroot (*Beta vulgaris*), Swiss Chard (*Beta cicla*)

Growing: Beets are incredibly nutritious and can be eaten raw or cooked. While the colorful root bulb is the more common food, some people use the beet greens in salad from time to time. Don't eat too much of those, as they can make you lose calcium in your body. Beets are cold hardy and need fertile soil. Harvest when they reach at least 1.5 inches (3.8 cm) in diameter. They can be stored in live storage or pickled. Swiss Chard is in the same family but doesn't form a fleshy root. Instead, the leaves are harvested as they grow, like lettuce. These highly nutritious greens can be eaten raw when young, but it is more common to cook them like spinach. Both of these varieties can be grown in the greenhouse over the winter.

Bishop's Weed

Uses: ground cover, edible

Species: Bishop's Weed (*Aegopodium podagraria*)

Growing: Bishop's Weed, or Goutweed, is a perennial plant that lives in most of the cooler areas of North America. In the wrong place, it is aggressive and tends to displace other species, which is why people do not like it. This tendency to spread and persist also makes it an excellent ground cover but creates too much work for many gardeners when they try to fight it by pulling it up and leaving the soil bare, which is still the perfect environment for it to grow. However, in the right spot it could be useful. If you have a distant

▼ Bishop's Weed

The Ultimate Guide to Permaculture

area of land that you don't use often but it needs cover, you can use Bishop's Weed as long as there is a deep and established plant border around it. As an edible green (for people as well), it can also be a useful forage crop.

Blackberry

Uses: edible fruit, bee forage
Species: Blackberry (*Rubus spp.*)
Growing: Blackberries grow wild everywhere in North America, and the sweet berries make excellent jam, desserts, and snacks. They are very hardy and easy to grow. With a trellis they can also act as a useful hedge. Simply pick the berries when they are ripe.

Black Locust

Uses: bare soil erosion control, bee forage, wood and timber, windbreak, dryland tolerant, leguminous nitrogen fixer
Species: Black Locust (*Robinia pseudocacia*)
Growing: Fairly hardy tree that prefers well-drained soil. Grows about 50–75 feet high and 35–50 feet wide. It usually grows in the eastern United States, although in some places this tree can spread and is considered invasive, so double-check your local species. In the first couple of years the seedling is particularly tasty to deer, and so it needs to be protected. It is considered one of the most valuable bee forages in France and makes very tasty honey. The wood is hard and durable, and it it resists rot. As a source of firewood it grows quickly and produces a long-burning log that makes very little smoke. The seedpods are toxic to animals when eaten.

Black Tupelo

Uses: bee forage, bare soil erosion control, stream bank erosion control, wood and timber, wetland tolerant
Species: Black Tupelo (*Nyssa sylvatica*)
Growing: This tree is native to the eastern United States and is also called the Blackgum. Besides the value to bees, the wood is hard and heavy. It is difficult to split, which makes it a good choice for wood items that take a heavy beating, like bowls or pulleys. It enjoys partial shade and can grow in many soil types, including marshy areas.

Blueberry

Uses: edible fruit, bee forage, ground cover
Species: Highbush Blueberry (*Vaccinium corymbosum*), Rabbiteye Blueberry (*Vaccinium asbei*), Lowbush Blueberry (*Vaccinium angustifolium*), Creeping Blueberry (*Vaccinium crassifolium*)
Growing: Blueberries enjoy full sun and grow to be 6 to 12 feet (2–4 meters) high and 6 to 12 feet wide. They need highly acidic, well-drained soil. The Highbush varieties are hardier and also more common, but the Rabbiteyes produce more fruit. Lowbushes don't make the greatest groundcover like the other varieties, but they have the best fruit. The fruit can be eaten raw, sun-dried, or as dessert and jam, and it is very nutritious. Blueberries do better with regular watering. The picking season starts sometime around May or June and lasts until the end of summer.

Borage

Uses: bee forage, edible

Species: Common Borage (*Borago officinalis*)

Growing: Borage is a valuable medicinal plant and is also a nutritious addition to salads. It also grows well as a companion plant. It is a hardy herb and, blooms most of the season, and re-seeds itself. The leaves are picked as needed.

Buckwheat

Uses: bee forage, edible, cover crop

Species: Common Buckwheat (*Fagopyrum esculentum*)

Growing: Although buckwheat is eaten as a grain, it is not a grass or a cereal. It grows very quickly, ripening in about 10–13 week and thus making a great crop for a cooler climate. It does not tolerate frost and should be planted late in the season to ripen in early September. It requires regular watering and soil that does not have to be great but at least loose. Because of its short growing season, it should be double-cropped with another grain, such as winter wheat, oats, or flax. If your earlier grain crops failed, it may also be possible to plant buckwheat and quickly raise an emergency food source. As a cover crop it effectively chokes out weeds. The seeds have a hard outer hull, which must be removed, and the seed can be used whole or ground as flour. Traditionally, buckwheat has been used for noodles and pancakes. It has no gluten and so can only be used in small amounts when making bread. The hulls are used as a filling for pillows or other items. To harvest, cut the buckwheat down when it is almost all brown but still has a couple of green leaves or flowers. Cut it as you would a grain, by mowing and swathing. Thresh the grain, remove the chaff, and then allow the seeds to dry. The seeds still have hard hulls on them, but they need to be very dry to be removed (and you don't want them to get moldy). You can do this by laying them out on your drying trays in the hot sun or using your dehydrator. Then you can use a grain mill to de-hull the seeds. You must use the largest setting and run them through a few times to crack open all of the hulls. Sift them until you have removed all of the hulls, which you can save for pillows or other projects.

Bunchberry

Uses: streambank erosion control, wetland tolerant, edible fruit, ground cover

Species: Canadian Bunchberry (*Cornus canadensis*), Bunchberry (*Cornus suecica*)

Growing: Bunchberries are perennials that like shady, cool, moist places. They are found in northern areas, growing around stumps and mossy areas that stay wet most of the time. Bunchberries need to be in a spot where the soil never gets warmer than 65°F (18°C), and they hardly ever see the sun. They enjoy soil full of rotting wood, which can be added as mulch. The berries are edible, although a little bland, and are best used in sauce, jams, and pudding. The berries also help thicken up jam, so you don't need to add pectin. The berry tea is a traditional herbal remedy.

Carob

Uses: seedpod and sugar source, edible, dryland tolerant, leguminous nitrogen fixer

Species: Carob Tree (*Ceratonia siliqua*)

Growing: Carob is a large evergreen shrub that grows to 50 feet (15.2 meters) and prefers hot climates. In the Mediterranean and Australia it is grown commercially, but in North America it could be grown in the southern states for its fruit as well. It is not frost hardy, although a full-grown tree could withstand 20°F (7°C) if the low temperatures do not occur during blooming, or otherwise they would prevent any fruit from growing. It won't tolerate any kind of heavy rain until the end of the growing season. It can grow in any soil and is tolerant to drought. The carob pods are sweet and have many uses. To harvest the pods, shake the tree with a long pole before the winter rains start. Catch the pods on a sheet or tarp on the ground. Allow the pods to dry in the sun for a couple of days until the seeds rattle in the pod. The pods can be crushed to break up the seeds and fed to animals, and for people they are most often ground into varying consistencies for use in making syrup, jam, fine flour, and a coffee substitute. Animals should not eat too much of the crushed pods, as their growth will be stunted. Chickens cannot eat them at all.

Carrot

Uses: edible roots and shoots

Species: Carrot (*Daucus carota*)

Growing: Carrots are a familiar root vegetable that comes in orange, white, purple, red, and yellow. The greens are also edible. They are a useful companion plant, especially for tomatoes. Carrots enjoys partial shade to full sun in loose soil. They take at least 4 months to grow.

Cattail

Uses: water plants, edible

Species: Narrowleaf Cattail (*Typha angustifolia*), Southern Cattail (*Typha domingensis*), Broadleaf Cattail (*Typha latifolia*)

Growing: All parts of the cattail are edible. The young shoots are cut in the spring when they are at least 4 inches (10 cm) long, and the roots can be boiled like potatoes, the flower stalks can be boiled or steamed like corn, and the pollen works as a flour substitute. They grow everywhere in the world, and the cattails can also be woven into useful items like mats and baskets. They are perennial water plants and unless kept in check can take over a pond. They are hardy and easy to grow from cuttings or seeds.

Ceanothus

Uses: leguminous nitrogen fixer, ground cover, edible

Species: New Jersey Tea (*Ceanothus americanus*), Snowbrush Ceanothus (*Ceanothus velutinus*), Maritime Ceanothus (*Ceanothus maritimus*)

Growing: Ceanothus is an evergreen species, but in very cold climates a few species are deciduous. The leaves are edible and have traditionally been used in herbal medicine and teas. There are many varieties of Ceanothus, all very hardy, and so it is probably a good idea to just pick your native species.

Cedar

Uses: pest control, wood and timber

Species: Deodar Cedar (*Cedrus deodara*), Lebanon Cedar (*Cedrus libani*),

Cyprus Cedar (*Cedrus brevifolia*),
Atlast Cedar (*Cedrus atlantica*)

Growing: Cedars grow in the mountains
where the winters aren't too severe
but still get snow or monsoon rains.
They grow very big and should not
be planted near any overhanging
obstruction or over buried pipes or
cables. They enjoy partial shade to full
sun. The wood is resistant to rot and
pests and has a pleasant scent.

Cherry

Uses: edible fruit, wood and timber, bee
forage

Bush Species: Mongolian Bush Cherry
(*Prunus fruticosa*), Japanese Bush
Cherry (*Prunus japonica*), Nanking
Cherry (*Prunus tomentosa*), Choke
Cherry (*Prunus virginiana*)

Growing: There are hundreds of cherry
varieties, but the bush varieties may
be the most useful as they don't take
up much space, growing 3 to 8 feet
(1–2.5 meters) high and up to 8 feet
wide. There are even dwarf bush
varieties that take up even less space
and are even more resistant to pests
and diseases. Mongolians are hardy
but sourer. Nankings are sweeter
and grow anywhere, including dry
places. Cherries enjoy long, warm
summers but need to get cold in the
winter, which makes them perfect for
a temperate climate. They need fertile,
well-drained soil. Sour cherries are
often self-pollinating, but sweet cherries
need to be planted in groups. Pick the
cherries as they ripen, eat them raw,
dry them, or make them into pies.

Chestnut

Uses: edible nut, wood and timber

Species: Chinese Chestnut (*Castanea
mollissima*)

Growing: The Chinese Chestnut grows to
be 80 feet (24 meters) tall and is self-
pollinating. It enjoys well-drained soil of
any type, is tolerant to cold, and prefers
partial shade to full sun. The chestnut
can be eaten raw, but it is much better
cooked and can be used as a staple like
potatoes. When the nuts are ready to
harvest, the spiny burs will begin falling
to the ground, which usually happens
in autumn. Most of the burs will split
open. Gather those up and remove the
nuts, throwing out any that are damaged
or have holes in them. Put them in an
airtight container or freeze them.

Chicory

Uses: edible roots and shoots, bee forage

Species: Chicory (*Cichorium intybus*)

Growing: Chicory tastes a little bitter, but is
very nutritious. It goes well with other
baby greens in salads. It enjoys partial
shade to full sun, and some varieties
are perennial. They are often pulled
out of yards as a weed because they
can take care of themselves very well.
They make excellent animal forage,
as the chicory kills worms and is
easy to digest. It also makes a useful
herbal remedy and coffee substitute.
Radicchio and Belgian endive are in
the same family.

Clover

Uses: bee forage, ground cover, edible

Species: White Clover (*Trifolium repens*),
Bush Clover (*Lespedeza bicolor*), Red
Clover (*Trifolium pretense*)

Growing: Clover enjoys partial shade to
full sun and is a helpful nitrogen fixing
ground cover. It is also a valuable

animal fodder and very hardy to stomping on. Clover is easy to grow and spreads to form a mat.

Comfrey

Uses: bee forage

Species: Large-flowered Comfrey (*Symphytum grandiflorum*), Russian Comfrey (*Symphytum x uplandicum*)

Growing: Comfrey is easy to grow in most climates, with the option to cut it back as mulch a few times per year. Once you have decided to grow it, you will never be able to get rid of it, so be sure you want it. It is mentioned several times in this book for its use as a fertilizer, because it adds many nutrients to the soil. It prefers lots of nitrogen, and so a bed of manure is a great home for it. It is useful as a topical medicinal herb but can cause liver damage when ingested in large quantities.

Cranberry

Uses: water plants, edible fruit, bee forage

Species: Common Cranberry (*Vaccinium oxycoccos*), Small Cranberry (*Vaccinum microcarpum*), Large Cranberry (*Vaccinium macrocarpon*), Southern Mountain Cranberry (*Oycoccus erythocarupus*)

Growing: Cranberries are evergreen dwarf shrubs and vines that grow sour red berries. The cranberries can be made into sauce, juice, jams, and dried fruit by adding sweeteners to make them delicious. They are grown similarly to rice, in a paddy surrounded by a barrier. While the plants require regular irrigation, the field is not flooded until the end of the year. While big producers float the berries to harvest them, small producers will want to dry pick the berries to avoid damaging the crop and use less

▾ Comfrey

equipment. Then the beds are flooded so that they freeze over in the winter. Every 4 years, a thin layer of sand can be spread on the top to help control pests. The cranberries are stored in well-ventilated crates in a cool dark place until they are processed.

Currant

Uses: edible fruit
Species: Black Currant (*Ribes nigrum*), Clove Currant (*Ribes odoratum*), Red and White Currant (*Ribes Silvestre*)
Growing: In Europe these are as common as blueberries, but in many places in North America they are illegal to grow because some species harbor a pest that kills white pines. They are very hardy to cold temperatures and can grow in most climates. Their one weakness is a susceptibility to rust, but many varieties are resistant, especially Red and White Currants. They are generally easy to grow, but check on your state's restrictions.

Dandelion

Uses: edible plant, bee forage
Species: Dandelion (*Taraxacum officinale*)
Growing: Dandelions grow spontaneously on their own in everyone's yard and don't need any care at all. They are also extremely nutritious and the entire plant can be eaten. The flowers are made into wine or added to pancakes, the roots are brewed into a tea that tastes like coffee, and the leaves are added to salads.

Daisy

Uses: pest control
Species: Pyretheum Daisy (*Chrysanthemum cinerariifolium*)

Growing: These flowers prefer dry and somewhat sandy soils. They need to be weeded, but since they are so hardy against pests, they should not have many other problems. To harvest them for use as a natural pesticide, wait for a warm, sunny day when the flowers have been already open for a few days. Dry them by hanging upside down or removing the heads and drying them in the sun. The flowers must be stored whole in a dark, airtight container. When you are ready to use them, grind them into a fine powder and either dust or spray them (mixed with some water) on the plants that are affected. The insects should die almost immediately, without harm to humans.

Daylily

Uses: bare soil erosion control, edible roots and shoots, bee forage
Species: Tawney Daylily (*Hemerocallis fulva*)
Growing: Daylilies enjoy partial shade to full sun. The entire plant can be eaten, including the flowers and roots. They are a little too large and expansive to be grown near other plants, but since they require little care, they can be planted in the outer zones. The Tawney Daylily is wetland tolerant. The edible parts of the daylily are the flower buds and flowers, which are most often used in Asian stir-fry dishes. Cut off the base with the ovary and use like mushrooms. Only the cultivated varieties of daylily are edible; others are toxic, so make sure you have the right type.

Dill

Uses: pest control, edible
Species: Dill (*Anethum graveolens*)

Growing: A perennial herb that enjoys full sun. It grows in just about any climate but prefers fertile, well-drained soil. The leaves can be used fresh, and the dill seeds can be harvested by cutting off the flower heads when the seeds begin to ripen. Place these flower heads upside down in a paper bag and allow them to dry in a warm, dry place. After 1 week, remove the seeds and store in an airtight jar to use in flavoring foods.

Duck Potato

Uses: water plants, edible roots and shoots
Species: Duck Potato (*Sagittaria latifolia*)
Growing: Also known as Broadleaf Arrowheads or Wapato. These produce a root vegetable that is eaten like potatoes. They grow directly in water or in marshy soil and can be planted from eyes just like potatoes. When harvesting, no more than ¼ of them should be collected per year, and they will spread to replace the loss. Gather these anytime during the summer and fall until the first frost.

Duckweed

Uses: water plants, edible
Species: Common Duckweed (*Lemna minor*), Star Duckweed (*Lemna trisulca*), Swollen Duckweed (*Lemna gibba*)
Growing: This versatile plant looks less like a plant than a small green disk that just floats on the surface of water. It is very high in protein, containing even more than soybeans, and although it is commonly used to feed water livestock like fish and ducks, it is edible and eaten by people in some places in Asia. Once introduced to a freshwater environment, it spreads quickly and covers the entire surface, unless an animal lives there to eat it. On small ponds, Duckweed helps to prevent evaporation and also provides shade to small fish.

Fennel

Uses: pest control, edible
Species: Sweet Fennel (*Foeniculum vulgare*)
Growing: Fennel is easy to grow, and in fact because of how quickly it spreads, it is now on invasive species lists in North America. The dried seeds are used as a spice, and the root bulb and leaves are used raw or cooked. Fennel has many medicinal uses. Be aware that Poison Hemlock looks very similar to fennel and grows near water or wet soil. Crush the leaves and smell them. Fennel has the distinctive smell of licorice, and Hemlock smells musty.

Fenugreek

Uses: edible, nitrogen fixing legume
Species: Fenugreek (*Trigonella corniculata*)
Growing: Fenugreek is an annual herb that enjoys full sun and well-drained soil. The seeds are the most useful part and can be harvested in early fall when the pods have dried. Store them in an airtight container. These are a useful herbal remedy and are also used in Indian food for making pickles, curry, and sauces. The leaves can also be used in salads, although they have a particularly bitter taste.

Fig

Uses: edible fruit, dryland tolerant
Species: Common Fig (*Ficus carica*)
Growing: Fig trees grow about 20–30 feet (6- 9 meters) tall and can grow in most places that experience a long

hot summer. They enjoy full sun and regular watering. They are one of the oldest and more nutritious human food crops, being high in calcium. The fruit can be eaten fresh, dried, or as jam. Use caution when handling the tree, as the sap is a skin irritant. The tree produces two crops per years, one in the spring and one in late summer or fall. The first crop, or *breva*, is usually very small, but a few varieties do produce a little more breva fruit.

Flowering Dogwood

Uses: wetland tolerant, ground cover, edible, wood and timber

Species: Flowering Dogwood (*Cornus florida*)

Growing: This variety of Dogwood grows mostly in the eastern United States and doesn't get much taller than 40 feet (12 meters). The fruit of this tree is poisonous to humans, but the bark is used in traditional medicine. Wildlife and birds eat the seeds, and the leaves make valuable forage for grazing animals. Dogwood is not very hardy, can't handle dry periods or sudden temperature changes, and is susceptible to pests. It enjoys rich soil , regular watering, and partial shade.

Gliricidia

Uses: bee forage, wood and timber, edible leaves and pods for animal fodder, nitrogen fixing legume

Species: Gliricidia (*Gliricidia sepium*)

Growing: Gliricidia enjoys shade and warm temperatures. The leaves, shoots, and pods make a very high quality protein-filled source of cattle fodder, and the tree itself is useful for live fencing, firewood, and green manure. It is the second most useful leguminous tree, after Leucaena. It grows very quickly to a height of 10

▾ Flowering Dogwood

The Ultimate Guide to Permaculture

▲ Gooseberry

feet (3 meters), and because of this quick-growing characteristic it is often used as a shade plant for other gardens. Use caution, however, as the leaves are great for cattle but toxic to horses and other livestock.

Gooseberry

Uses: bee forage, edible fruit
Species: Gooseberry (*Ribes uva-crispa*)
Growing: Gooseberry enjoys partial sun to full shade and grows 3 to 5 feet (1–1.5 meters) tall and wide. In some places gooseberries are illegal as they sometimes harbor a pest that kills white pines, so be aware. They are very hardy and grow in most climates.

Grape

Uses: edible fruit
Species: Fox Grape (*Vitis labrusca*), Muscadine Grape (*Vitis rotundifolia*)

Growing: Unlike human-cultivated species of grapes, Foxes and Muscadines are much hardier against pests and diseases. They are best grown on a trellis, which requires more effort in pruning but is worth the trouble. The trellis needs to be extremely strong because in a few years the grape will grow to be very heavy and tree-like. Foxes are self-pollinating, but Muscadines need several plants to make sure there are both male and female varieties. Foxes grow in any climate, and the Muscadines need the warm, humid climate found in the south. Grapes can be eaten raw, or made into jam, juice, vinegar, wine, raisins, oils, or seed extracts. While grapes grow anywhere, they don't produce fruit unless they have a long, hot growing season, which is challenging in a northern climate. They enjoy full sun and the heat of the south side of a house, and they need regular watering in well-drained soil. It may not be worth growing them in cold climates, as the entire vine must be laid flat on the ground and mulched, and the buds must be protected from frost every year.

Groundnut

Uses: edible roots and shoots, nitrogen fixing legume

Species: Groundnut (*Apios americana*), Fortune's Groundnut (*Apios fortune*), Price's Groundnut (*Apios priceana*)

Growing: A groundnut is somewhat like a "new" potato, of similar size, and with lots of little roots coming off it, but unlike a potato it is full of protein. It climbs and sends out roots, invading the neighboring areas, and so it must be kept in check. The roots can be cooked exactly like any other root vegetable.

Guava

Uses: edible fruit

Species: Apple Guava (*Psidium guajava*)

Growing: While the guava is a subtropical plant, it is surprisingly hardy and can survive temperatures as low as 40°F (4°C) and very little rain. As long as your climate does not experience frost and you protect young plants, you can grow guava. Guava plants can bear fruit even when grown in containers and usually do so within a couple of years. The fruit is eaten raw or cooked, or it is made into juice. Guava plants do better with regular watering that reaches their deep roots, rather then allowing their roots to dry completely before watering again. They need fertile, well-drained soil and plenty of nitrogen in the form of manure. When the fruit is ripe, it will change in color and smell dramatically. Guavas are best when allowed to ripen on the tree,

▾ Guava

The Ultimate Guide to Permaculture

but green fruit can be stored for up to 5 weeks in cold storage.

Hawthorn

Uses: edible fruit, bee forage, erosion control, windbreak, dryland tolerant

Species: Hawthorn (*Crataegus L.*)

Growing: There are many varieties of Hawthorn, which is a small tree growing around 20 feet (6 meters) high. It enjoys partial shade to full sun and well-drained soil. The berries are edible and are usually used cooked like apples, in jams and pies.

Hazelnut

Uses: edible nut

Species: Common Hazel (*Corylus avellana*)

Growing: The Hazel is a shrub that grows about 10–20 feet (3–6 meters) high. Traditionally, besides producing a protein-filled nut, it also serves the useful purposes of being part of a hedge, and also as a fencing material. The flexible branches can also be made into plant supports and arches for climbing species. Hazel shrubs prefer well-drained soil but do better in rainy areas, especially in the Pacific Northwest. Harvest the hazelnuts in mid-fall when the nuts and leaves begin to drop. Allow the nuts to dry, and then eat them raw, roasted, or ground into a paste.

Hickory

Uses: wood and timber, edible nuts

Species: Northern Pecan (*Carya illinoinensis*), Shellbark Hickory (*Carya laciniosa*), Shagbark Hickory (*Carya ovate*)

Growing: Hickories enjoy full sun and grow about 70–100 feet (21–31 meters) tall. The Northern Pecan gets to be 75–120 feet (23–37 meters) wide, while the hickories are 30–50 feet (9–15 meters) wide. Hickories need other hickories nearby to pollinate, and so you will need to plant a few. It is recommended to graft pecans onto other hickories. The nut will be improved, and the pecans will bear sooner than the usual 10–15 years. The hickories and pecan species listed here combined cover all areas of North America, but you need to pick the right species for your climate. Soil types don't affect the hickory family much, but temperature does. The hickories are more valuable for their wood, which is used in smoking meat and makes excellent firewood. Hickory also makes very durable furniture. The hickory nut is just as good as pecans, and these are all harvested as they start to fall to the ground. Shake off the rest. Hull them, wash the nuts, and spread them out to dry. The drying process in a well-ventilated room or garage with a fan takes about 2 weeks.

Honey Locust

Uses: edible seedpod, dryland tolerant, nitrogen fixing

Species: Honey Locust (*Gleditsia triacanthos*)

Growing: The Honey Locust can reach a height of 60–100 feet (18–31 meters) and lives to be about 120 years old. It is sometimes described as a perennial shrub rather than a tree. It prefers well-drained fertile soil. The pulp inside the seedpod is edible (although be careful as the Black Locust is not), and the pods make great forage for grazing animals. The tree is hardy, grows

▲ Honey Locust

quickly, and provides massive areas of shade for people and animals.

Honeysuckle

Uses: wetland tolerant, edible.
Species: Blueberried Honeysuckle (*Lonicera caerulea*)
Growing: There are many types of honeysuckles, but only few have edible berries. Russia has been raising Blueberried Honeysuckles as a food crop for a long time. They prefer wet and marshy soil and are usually found in the Northeast. Honeysuckle has never been found wild in the Pacific

Northwest. The fruit can be used for jam, juice, wine, ice cream, yogurt, and sauces. It can handle very cold temperatures and needs to have two compatible varieties in order to produce fruit. Harvest the fruit as it ripens.

Hops

Uses: edible
Species: Common Hop (*Humulus lupus*)
Growing: Hops are a perennial climbing plant that produces flowers called hops, which are used mostly in beer production. The extract from the hops is antimicrobial and is naturally sedative. Hops take up a lot of space and enjoy full sunlight. They need regular watering and well-drained soil that is rich in minerals and nitrogen. It takes about 4 months for them to produce flowers, but they grow well in most places as long as they are not allowed to freeze. They are susceptible to molds and insects, which must be prevented with a drip irrigation system and organic pesticides. The flowers are ready to harvest at the end of summer when the flower cone has become light and dry and doesn't stay compressed when squeezed. Dry these cones on a drying rack in the sun or in a dehydrator and stir and rotate them daily until the last cone is springy and the powder falls out easily. Seal them in a freezer bag and put them in a freezer until you are ready to use them.

Horseradish

Uses: edible
Species: Horseradish (*Amoracia rusticana*)
Growing: Horseradish is related to cabbages and broccoli, and as such is a hardy and strong-flavored perennial

▲ **Hops on trellises.**

that grows just about anywhere. It prefers a long growing season and a cold winter. The root is harvested after the first frost. A few pieces are cut off for replanting, while the rest is kept for yourself. If it is left in the ground, it will shoot out sprouts underground and take over the garden. Horseradish is a great flavoring for food but do not eat it raw in any great quantity, as it will cause digestive upsets. It must be grated and added to vinegar, where it will keep for several months. When it begins to darken in color, it is time to throw it out. Horseradish sauce is simply the vinegar combination mixed with mayonnaise. It is also a valuable herbal medicine.

Hyssop

Uses: bee forage, edible
Species: Herb Hyssop (*Hyssopus officinalis*)

Growing: Hyssop is a hardy perennial that stays short and can make a tiny hedge. It prefers full sun and well-drained soil. Hyssop has a slightly minty flavor and can be added to soups, but it is far more valuable as a medicinal herb for which it has hundreds of uses. Harvest the plants just before the flowers begin to open, and hang upside down to dry. Remove the leaves and flowers and place them in an airtight container.

Japanese Pagoda Tree

Uses: nitrogen fixing legume
Species: Japanese Pagoda Tree (*Sophora japonica*)
Growing: A Japanese Pagoda tree grows about 50–75 feet (15–23 meters) high and 50–75 feet tall. It enjoys full sun and likes fertile, well-drained soil. It is often planted in smoggy cities, as it is very hardy to pollution. It needs regular

watering. The leaves and flowers have traditionally been used in herbal medicine, but their bitterness keeps them from being eaten regularly. The birds, however, are very attracted to the seedpods.

Jasmine

Uses: wetland tolerant, edible, climbing vine

Species: Jasmine (*Jasminum L.*)

Growing: There are over 200 different types of Jasmine and they grow all over the world. In Asia they are grown for their flowers, which only open in the evening and are useful not only for their beauty, but as a flavoring for green tea. The flowers can also be used to make extracts of syrup and essential oil. There are two kinds of Jasmine, one that stays green all year, and another that loses its leaves in the winter. They enjoy warm, moist climates, and regular watering.

Jerusalem Artichoke

Uses: edible roots

Species: Jerusalem Artichoke (*Helianthus tuberosus*)

Growing: These produce an edible root vegetable that looks a little like a yam. Each plant produces tons of food and is likely to spread unless contained. Jerusalem Artichokes enjoy partial shade to full sun and can be grown in most climates. They are a perennial and not related to artichokes at all, but are actually closer to sunflowers. Because of the quantity of food produced by each plant and the high sugar carbohydrate content, Jerusalem Artichoke is a good source of fructose and can be made into ethanol and yeast for fermentation. The plants are

▾ **Jerusalem Artichoke**

The Ultimate Guide to Permaculture

▲ Jicama

easy to grow, but every year you must dig them up and replant them in order to prevent them from taking over. They can be eaten cooked like potatoes or as a substitute for turnips and parsnips, and they make excellent animal fodder, especially for pigs, which can dig them up themselves. Jerusalem Artichokes will cook much faster than potatoes and so must be watched so they don't turn into mush.

Jicama

Uses: edible roots and shoots
Species: Jicama (*Pachyrhizus erosus*)
Growing: Also known as Mexican turnips or Yam Beans, the Jicama (with the "J" pronounced as an "H" sound) has a tasty root resembling a potato and is very sweet. It is usually eaten raw and

is sometimes used in stir-fry. It is can be prepared with a variety of spices and is added to salads or dipped in salsa. Be cautious, however: The rest of the plant is highly poisonous. Only the root may be eaten. It is not the most nutritious of foods but is full of fiber and has sugars that are beneficial to the friendly bacteria of the digestive system.

Jujube

Uses: edible fruit, wood and timber, dryland tolerant
Species: Common Jujube (*Ziziphus zizyphus*)
Growing: Jujube grows in many climates and conditions and is very hardy, although it needs to have a long hot summer and regular water to produce

fruit. It can survive cold winters and extremely hot days in the summer. The fruit is edible and used in traditional medicine, as well as eaten raw or made into candied dried fruits and jams.

Kiwifruit

Uses: edible fruit

Species: Hardy Kiwifruit (*Aetinidia arguta*), Super-Hardy Kiwifruit (*Actinidia kolomikta*), Purple Hardy Kiwifruit (*Actinidia purpurea*)

Growing: These hardy kiwi species enjoy full sun but can be grown almost anywhere. Unlike the grocery store variety, they are much smaller and less fuzzy. They also taste sweeter. You will need to plant male and female trees for pollination and protect them from frost when they are young, but once established, they will grow prolifically and provide hundreds of pounds of fruit every year. They will grow very tall, and the fruit may be difficult to reach unless you train it to a trellis, which requires much more effort in pruning but is worth it. The fruit can be stored like apples in the cold storage.

Lavender

Uses: bee forage, dryland tolerant

Species: English Lavender (*Lavendula angustifolia*)

Growing: Lavender enjoys full sun and well-drained or even sandy soil. If it is too moist or too well fertilized, lavender can be susceptible to mold. The flowers are used to extract essential oil, which is a popular natural remedy, and they are also edible and sometimes candied, added to teas, made into syrup and dried. For most uses the flowers are harvested when they are open and at their brightest. Cut the stems when they are dry and cool, especially in the morning after the dew has dried. You can either hang the flowers upside down or remove the heads and spread them on a drying rack in the sun or in the dehydrator.

Lemongrass

Uses: streambank erosion control, edible

Species: Lemongrass (*Cymbopogon citratus*)

Growing: Lemongrass is a perennial herb that makes a grassy clump about 3–5 feet (1–1.5 meters) tall. It prefers full sun and rich, moist soil. You can fertilize monthly during the summer. Cats are the plant's only real enemy, as they eat the leaves. Lemongrass is a common ingredient in Asian cooking and is used in tea, soup, curry, and meats either fresh or powdered. Harvest the leaves any time after the plant is over a foot tall, and dry or use fresh.

Lespedeza

(see Clover)

Leucaena

Uses: streambank erosion control, ground cover, wood and timber, edible, nitrogen fixing legume

Species: Leadtree (*Leucana benth*)

Growing: The Leucaena species is rated as the most useful nitrogen-fixing legume. It is native to the American continents, and there are 24 species of trees and shrubs from all different areas and climates, although they prefers warm climates and regular watering. Leucaenas do not tolerate frost, although frost won't completely kill them. The seedlings must be protected from weeds and animals, but once established, they are fairly hardy.Leucaena is an excellent animal forage and a quick-growing erosion control plant. Its leaves are a green manure, and its wood is an efficient firewood.

Licorice

Uses: nitrogen fixing legume, edible

Species: Licorice (*Glycyrrhiza glabra*), American Licorice (*Glycyrrhiza lepidota*), Chinese Licorice (*Glycyrrhiza uralensis*)

Growing: Licorice is a perennial herb that enjoys full sun and well-drained soil. It takes a couple of years before it can be harvested. To harvest, the root is boiled to extract the licorice flavor, and then evaporated to concentrate it. The concentrate can be made into a dry powder or syrup that is much sweeter

▾ **Loquat**

than sucrose. It has many medicinal uses, but excessive use can cause liver and heart problems. Eating it too frequently can cause a rise in blood pressure.

Loganberry

Uses: bee forage, edible

Species: Loganberry (*Rubus loganbaccus*)

Growing: Loganberries are hardy to pests, disease, and frost. They grow as a bramble rather than tall canes (like raspberries). Individual canes die after a couple of years and should be cut off. The fruit ripens early and produces for about 2 months. It should be picked when it has changed from a deep red to a deep purple color. Loganberries can be eaten raw or make into jams and dessert. Use them exactly like raspberries and blackberries.

Loquat

Uses: edible fruit

Species: Loquat (*Eriobotrya japonica*)

Growing: The Loquat tree is usually small and stays around 10 feet (3 meters) high. It prefers warm temperatures, full sun, and well-drained soil. It is fairly hardy, doing well in any kind of soil, but needs regular watering and warmth. The Loquat fruit is similar to apples and can be eaten raw or made into jam, jelly, and chutney. The leaves are edible and often used to make a nutritious and medicinal tea.

Lotus

Uses: water plants, edible

Species: American Lotus (*Nelumbo lutea*), Sacred Lotus (*Nelumbo nucifera*)

Growing: There are many varieties of lotus plants, some more cold hardy than others, so double check the species you choose for their minimum lowest temperature. They need full sun all day and are planted in water, which means that your pond cannot be shaded by any trees. They are usually planted as tubers or root bulbs that are pushed into the soil in the shallow end of the pond and weighted down with a rock. They need lots of fertilizer, so they do well with fish, but large fish will eat them very quickly. These tubers are edible to humans as well and can be cut up and used in stir-fry.

Lucerne

(see Alfalfa)

Lupine

Uses: bee forage, edible, nitrogen fixing legume, ground cover

Species: Wild Lupine (*Lupinus perennis*), White Lupine (*Lupinus albus*), Yellow Lupine (*Lupinus luteus*), Blue Lupine (*Lupinus angustifolius*)

Growing: Lupine (or lupin) is a perennial herb that has all kinds of versatile uses. It has been found to have the equivalent nutritional values to soy and can grow in colder climates. The beans can be used as a grain and as a livestock feed. Make sure you and your animals do not eat the poisonous varieties, which can cause convulsions, liver damage, and even a coma. This includes Silky Lupine (*L. sericeus*), Tailcup Lupine (*L. caudatus*), Velvet Lupine (*L. leucophyllus*), Silvery Lupine (*L. argenteus*), and Lunara Lupine (*L. formosus*). Lupine grows in most climates and conditions and is very hardy, but it does take some work to make even the non-poisonous varieties

edible, unless you choose modern species that are specifically bred for this purpose. To harvest the beans, cut the plants down when the pods change from green to yellow and the lower leaves start to fall off. A cloudy day or in the early morning when the dew is still on is a good time. Let the beans dry in the shell on drying racks and thresh them like beans. Even the sweet varieties are so bitter as to be unpalatable and can make you sick if eaten raw. For humans, preparing the beans for consumption involves soaking the them for 24 hours, then simmering them for several hours. Then they must be put in a brine of very salty water for about a week in the cold storage or fridge, changing the water every day. Many people prepare them as pickles for this reason. After the brining process, chili, basil, and other spices can be added to taste. To feed animals with lupine, you don't have to soak it, but you will have to grind it up and mix it with other feed crops to disguise it. You might have to add a little molasses as well.

Macadamia

Uses: edible nut

Species: Macadamia Nut (*Macadamia integrifolia*), Prickly Macadamia Nut (*Macadamia tetraphylla*)

Growing: Macadamia trees are medium-sized evergreen trees that produce a delicious edible nut. They won't produce nuts until they are at least 7 years old. They enjoy fertile, well-drained soil, warm temperatures, and regular watering. The roots are very shallow, and the trees can be easily knocked down by strong winds. The nuts are harvested when they begin to fall off the tree, and they must be gathered every week during the harvest season. Dry the nuts on a drying frame they start rattling inside their shell. Once dry, they must be stored in an airtight container in the cold storage or the freezer. They can also be roasted for extra flavor.

Maple

Uses: source of sugar and edible pods, bee forage, wood and timber

Species: Sugar Maple (*Acer saccharum*), Box Elder (*Acer negundo*), Black Maple (*Acer nigrum*), Red Maple (*Acer rubrum*),

Growing: Maple trees enjoy shade to full sun and grows 75–100 feet (23–31

▾ Tapping maple syrup.

meters) high with a crown 75–100 feet wide. It takes 40 gallons (151 liters) of sap to make 1 gallon (3.8 liters) of syrup, so you will need at least a few trees, and while sugar maples produce more sap than other varieties, any species of maple can be tapped. They can be planted closely together and will grow in all but the sandiest soils. The one thing that does affect them is temperature. They prefer a temperate climate with a winter temperature of 0°–50°F (-18–10°C). Be aware that even though the sap is edible, the leaves are toxic, so it is better to keep animals and maples separate.

In March or April, when it the sap still freezes at night and thaws in the daytime, test to see if the sap is flowing by cutting a small gash in the side with the most limbs. Some will be running and some won't. When a tree is running, drive in a maple syrup spout and hang a bucket. The sap will drip into the bucket from the spout. Some trees produce more and tastier sap than others, and some years are just bad years so the whole process is out of your control. You will need to collect the sap in the morning and afternoon. Dump the sap into a very large pot or cauldron. Each tree will yield 6–12 gallons (23–45 liters). Heat the syrup up to low boil or simmer. You will have to stir it almost constantly to keep it from boiling over, so it is better to have more than one person take turns. As the sap boils down, keep adding more. The sap is done when its temperature is 219°F (103°C). Use a candy thermometer to measure this. Don't let it scorch, or it will ruin the syrup. A gallon will weigh about 11 pounds (5 kg). Stop tapping your trees when the snow melts, the ground thaws, and the buds start to swell. The sap just won't taste good. You can make sugar with the sap by letting it harden and grinding it up into a powder.

Marigold

Uses: pest control

Species: Mexican Marigold (*Tagetes cempasúhil*), African Marigolds (*Tagetes erecta*), French Marigold (*Tagetes patula*)

Growing: Marigolds enjoy full sun and most kinds of well-drained soil. Because of their pest control qualities, they are fairly hardy and don't need to be fertilized. They make excellent companion plants and are often used as a border plant as they can even deter deer and rabbits.

May Apple

Uses: edible fruit

Species: May Apple (*Podophyllum peltatum*)

Growing: Also known as American Mandrake, the May apple is a hardy perennial that enjoys well-drained soil and partial to full shade. Only the ripe fruit is safe to eat, and it does not ripen in May as its name suggests. Unripe fruit is toxic and causes severe digestive upset, and even ripe fruit can cause problems in large quantities. The seeds and peel are also not edible. However, once the skin has been peeled and the seeds removed, the ripe fruit can be eaten raw or used in jam or pie.

Mesquite

Uses: bee forage, dryland tolerant, nitrogen fixing legume, edible, wood and timber

▾ **May Apple**

Species: Honey Mesquite (*Prosopis glandulosa*), Velvet Mesquite (*Prosopis velutina*), Creeping Mesquite (*Prosopis strombulifera*), Screwbean Mesquite (*Prosopis pubescens*),

Growing: Mesquite is found wild throughout the southwestern United States and even up into the Midwest. It is a tree that can reach 20–30 feet (6–9 meters) high, although it generally stays small enough to look like a shrub. It grows a very long taproot that has been known to reach down as much as 60 feet (18 meters) to draw up groundwater, but it also draws up surface water if it needs to. Because of this, it can handle very long droughts and is extremely hardy. It grows and spreads quickly, and for this reason some people consider it a pest. The bean pods can be made into jam or dried and ground into delicious flour for bread. To harvest the pods, wait until the pods are beginning to fall from the tree. If they take more than a slight pull to come off the branch, they aren't ready yet. They should be as dry as possible. Rinse them and then lay them out in the sun on a drying rack until they snap in two when bent. To make flour, the whole pod is ground, but this would gum up a regular grain mill. The grinding can be done with a heavy-duty blender or a hammer mill. The meal or flour can be substituted for no more than half a cup per cup of grain flour. It is very nutritious. The pods also serve as animal forage, and the wood is hard and useful for furniture and other well-used items. The Honey Mesquite in particular is also used for decorative woodworking. Mesquite is commonly used as firewood and especially in barbecuing to add flavor to meat.

Mint

Uses: wetland tolerant, bee forage, ground cover, edible

Species: Apple Mint (*Mentha suaveolens*), Bowles Mint *(Mentha x villosa)*, Field Mint (*Mentha arvensis*), Corsican Mint (*Mentha requienii*), Peppermint (*Mentha piperita*)

Growing: This nice-smelling perennial herb enjoys partial shade and quickly spreads everywhere. It tends to prefer moist soil, although it will grow anywhere. Because of its spreading capacity, mint is often grown in containers to prevent it from taking over the garden. Apple and Bowles make great ground covers, however, and they also repel pests. The leaves are very healthy and can be used raw, dried, or in teas. These can be harvested throughout the season.

Moringa

Uses: seedpod and sugar source, edible

Species: Moringa (*Moringa oleifera*)

Growing: Moringa has huge potential to provide highly nutritious food. It prefers warm temperatures and well-drained soil, and it needs full sun. It requires regular watering. The tree grows very quickly and can reach 15 feet (5 meters) in 1 year, and it has the capacity to be cut down to 3 feet (1 meter) every year. It will grow back, and will produce edible pods that are within easy reach for picking. In several years it will produce thousands of pods. The leaves and pods can be used as animal forage without any processing, and they can also be harvested for human consumption. The leaves are used raw in salads, or dried and used as tea, or ground into powder and added to food for their high nutritional value. The flowers can be harvested and must be cooked and used like mushrooms. The pods can be used fresh raw or cooked in stir-fry and other dishes, or the seeds can be removed and used like peas.

Mulberry

Uses: edible fruit, dryland tolerant
Species: Red Mulberry (*Morus rubra*), White Mulberry (*Morus alba*)
Growing: Mulberries grow in most climates and will still produce fruit in partial shade. While many are somewhat tasteless, a variety can be chosen that has a stronger flavor, making them a very low-maintenance food provider. Mulberry tastes great raw or in a pie, but its real value lies in its high herbal and medicinal uses. The mulberry

leaves are also the sole source of food for silkworms, if spinning silk is of interest to you. Silk spinning is actually quite easy because the fibers are already long. Silk fiber is very durable and lightweight.

Myrtle

Uses: edible fruit
Species: Common Myrtle (*Myrtus communis*)
Growing: Myrtle is an evergreen shrub that is often used in topiaries. It can't handle extremely cold temperatures, but if it is planted near shelter will be fairly cold hardy. It prefers well-drained soil. The leaves have been used as herbal medicine for thousands of years, and the berries are edible, although they are very bitter.

Nasturtium

Uses: ground cover, edible
Species: Any

▾ Nasturtium

The Ultimate Guide to Permaculture

Growing: Nasturtiums are easy to grow and enjoy partial shade to full sun. The entire plant is edible, but people love to use the flowers since they are so brightly colored. They repel pests and are generally hardy. The unripe seedpods are often pickled to make a peppery condiment, or the ripe seeds can be dried and used exactly like peppercorns to replace pepper. The flowers and leaves are used raw or cooked to add a peppery flavor to any dish.

Natal Plum

Uses: edible fruit, ground cover
Species: Natal Plum (*Carissa macrocarpa*)
Growing: The Natal plum does not produce plums but is a shrub that produces a berries that are shaped like plums. The berries are very nutritious although tart. The plant can also be used as a hedge because of its thorny branches, without much care or effort, as it doesn't need much pruning, and dwarf varieties can provide a very effective ground cover. The plant grows in most conditions and doesn't need regular watering. It takes a few years before it will produce fruit. Pick the fruit when

▾ **Neem**

ripe and, be careful not to bruise it, and you can eat it raw, but it tastes better as a substitute for cranberries or in jam.

Narrow-leaved Tea-tree

Uses: pest control
Species: Narrow-leaved Tea-tree (*Melaleuca alternifolia*)
Growing: This plant produces the highly popular tea tree oil that is so heavily marketed today. It is native to the hot weather of Australia but has been grown with varying success in many other places. It enjoys full sun and only occasional watering, and it must be protected from frost when young. It is not an especially useful plant in the garden but is included here (like Neem) for the vast range of useful products that can be made from the oil. The leaves are picked, and the oil is distilled and bottled. It is antibacterial, antifungal, antiseptic, and antiviral. It is not edible but can be used for just about any skin problem. It is an extremely effective pest control.

Neem

Uses: pest control, dryland tolerant
Species: Neem (*Azadirachta indica*)
Growing: Neem is a sub-tropical evergreen tree that grows quickly. It is tolerant to drought and prefers well-drained soil. None of it is edible, and it has very little use in the garden, but it is included for the vast array of products that can be made from it. The entire tree has antifungal, antibacterial, sedative, and antiviral properties and is commonly used as a natural pesticide. The extracts have been used to make toothpaste, skin creams, sprays, and

a plethora of other items to cure just
about every type of skin ailment.

Nettle

Uses: edible, groundcover
Species: Stinging Nettle (*Urtica dioica*)
Growing: Nettles grow as either annuals
    or perennials and have stinging hairs
    on the leaves and stalks that greatly
    irritate the skin. Despite this, they are
    a very old medicinal plant which can
    be used for teas, shampoo, and skin
    salves. They are very hardy and grow
    very similarly to blackberries, which
    happen to be their direct competitor.
    Nettles are very high in vitamins and
    minerals, and it only takes soaking
    or cooking to remove the stinging.
    They are often used in polenta, pesto,
    soup, and with cooked greens. Nettle
    is also a substitute for flax or hemp in
making linen, and the processing is the
same. It deters pests and, if kept under
control, is a great companion plant.
Not all stinging varieties of nettles are
edible, so make sure you are eating
the right one. To harvest the leaves,
don a pair of gloves and use a pair of
scissors to cut the leaves off in March
or April before they begin to flower.
The smaller leaves near the top are the
best. Soak them in warm water for ten
minutes and drain off the water without
touching it. They are now ready to
cook.

New Zealand Spinach

Uses: dryland tolerant, edible, ground
    cover
Species: New Zealand Spinach (*Tetragonia
    tetraconioides*)

▾ Stinging nettle.

The Ultimate Guide to Permaculture

Growing: This leafy vegetable is very similar to spinach and is cooked like spinach. It prefers a hot, moist environment. Because of the way it spreads, it is considered invasive in some places, but this quality also makes it an excellent ground cover. It has no enemies, and even slugs don't like it.

Oak

Uses: wood and timber, edible acorn

Species: Holm Oak (*Quercus ilex*), Swamp White Oak (*Quercus bicolor*), Burr Oak (*Quercus macrocarpa*), Schuette's Oak (*Quercus x schuettei*), Cork Oak (*Quercus suber*)

Growing: All of the oaks enjoy full sun and grow 75–100 feet (23–31 meters) high and 75–100 feet wide. The Bur Oak is the hardiest and is suitable for the Midwest. It can grow some of the largest and sweetest acorns with less tannin. The Holm Oak prefers warmer climates but is dryland tolerant, making it suitable for southern regions. The Schuette is a hybrid that grows sweeter acorns, and the Swamp Oak likes moist areas. The Cork is an evergreen variety, and from it we get cork for wine bottles and other cork products. The bark is harvested without killing the tree, and it regrows in 150–200 years.

Acorns are not eaten fresh but have been processed and ground into flour for thousands of years, predating any flour made from grain. To do this, collect acorns when they are fully formed and beginning to fall, or off the ground. Don't take acorns without "hats" and throw out any with a hole in

them because they are likely to have a weevil living in them. Dry them in the oven, in a dehydrator, or in the sun, and then crack off the shell. The nutmeat is the part that you eat, but you must first remove the *tannins*, which are bad for you. Put the nutmeat into a big pot of water and bring to a boil. The water will turn dark as the tannin leaches out. Pour it off, fill the pot with water again, and boil again. Usually this takes at least 5 or 6 pots to remove all of the tannin. The alternative is to grind the nut into flour first and simply soak it in 1 part water to 3 parts flour. Keep that in the fridge for a week and switch out your water every day. Once the tannin is gone, dry the nuts again to prepare them for grinding. Grind them into a fine flour and use to replace up to half the flour in any recipe.

Olive

Uses: edible fruit, dryland tolerant

Species: Mission (*Olea L.*)

Growing: Olive trees enjoy the limestone soils near the ocean and prefer sandy and clay soil. They don't do well in fertile ground. They need hot weather and can't tolerate below 14°F (-10°C), but they are also very hardy to drought. The trees are susceptible to pests and fungus, which can be prevented with predatory wasps, pruning, and avoiding fertilizers. The olives are harvested in late fall when they begin to fall off the tree. For oil, the olives do not need to be handled gently, but for other uses they must be picked by hand. Mission olives do not need to be fermented before processing, like green olives do, but they do need to be cured. Wash the olives, allow them to dry thoroughly, and put them into a well-ventilated box lined with burlap or other cloth. Vegetable crates work well for this purpose. Mix in 1 pound (0.5 kg) of salt per 2 pounds (0.9 kg) of olives, and cover it all with another inch (2.54 cm) of salt. The boxes should be placed where they can drain, either on a waterproof tray or onto the ground. Leave them there for 1 week. Starting the second week you will have to mix them thoroughly every 3 days. Pouring them into another container and back again is an easy way to do this, and it's a good opportunity to pick out the soft or broken ones. Continue doing this every 3 days for 3 weeks, which may seem tedious but it's worth it. By this time, a month has now passed, and the olives should be shriveled up. Use a strainer to remove the salt and dip the olives for a few seconds into boiling water. Allow them to dry again completely and mix them with salt this time adding 1 pound (0.5 kg) of salt for every 10 pounds (4.5 kg) of olives. Put them into an airtight container and store in a cool place. In the cold storage they will last a month, but in the fridge or freezer they will last longer.

Onion

Uses: edible roots & shoots

Species: Multiplier Onions (*Allium cepa aggregatum*), Egyptian Walking Onion (*Allium cepa proliferum*), Nodding Wild Onion (*Allium cernuum*), Welsh Onion (*Allium fistulosum*), Chives (*Allium schoenoprasum*), Garlic Chives (*Allium tuberosum*)

Growing: All of the species above are perennials, which means that you don't need to replant every year. You can eat the entire plant, but to keep them going the next year, simply divide the clump in half and leave the rest in the ground. They grow incredibly easy, are cold and pest hardy, and will grow just about anywhere. The greens (or green onions) can be cut throughout the season, and the bulb is also harvested in the fall.

Palm

Uses: edible, wood and timber

Species: Coconut Palm (*Cocos nucifera*), Jucara Palm (*Euterpe edulis*), Açaí Palm (*Euterpe oleracea*), Peach Palm (*Bactris gasipaeas*), Cabbage Palm (*Sabal palmetto*)

Growing: There are hundreds of types of palms living in almost every climate, and yet many are endangered due to their usefulness as a source of timber. The problem is made worse by the palms' slow growth. Palms require partial shade to full sun, deep watering just after planting, and then no water unless they are extremely dry. The leaves gradually turn brown and fall off, but you can cut those off to keep the palms pretty. They will grow even in cold climates, as long as they are somewhat insulated in the winter. The Cabbage Palm is cold, fire, wind, and drought resistant. The Peach Palm is more of a warm-climate tree and produces an edible fruit that can be eaten raw or cooked. It is often made into jams or flour. The Açaí Palm also produces a fruit that is smaller than a grape that is extremely popular in Brazil and can be served with anything because of its amazing nutritional value. The leaves can be made into hats, baskets, and roof thatch, and the wood can be made into furniture. The Jucara is almost exclusively used for heart of palm, but all of these varieties

▾ Heart of palm.

can be harvested for heart of palm. Heart of palm is a delicacy, but for species that grow fruit, it does not make sense to cut down the whole tree to eat its heart. However, should you find yourself with a downed palm, then of course you should harvest the valuable heart. To do this, you must peel off the outer layers to reveal the white central core. This is very labor intensive. The core can be canned, fermented, or eaten fresh. The other edible product of palms is, of course, the nut of the Coconut Palm. The coconut you see in movies is actually only a small part of the coconut, which has a large husk around it. When coconuts are ripe, the husk will be a bright green color and must be cut off with a machete. The coconut can be processed into coconut oil for cooking, the meat can be eaten fresh or dried, the milk can be used raw or condensed, and the flour can be used in baking. Even the husk is useful in making charcoal, and the shell is used for cups and bowls all over the world. Be aware that the coconut is a common allergen and can cause food and contact allergies.

Palo Verde

Uses: seedpod, dryland tolerant, nitrogen fixing legume

Species: Blue Palo Verde (*Parkinsonia florida*), Yellow Palo Verde (*Parkinsonia aculeata*)

Growing: Palo Verde is a small tree with a green trunk that is difficult to grow from seed but once established is very hardy. It grows in dry, hot places, but it does need regular watering. It provides a shady canopy for other plants. Harvest the pods when they turn green, or let them dry on the tree. They are dry when they turn completely brown. Wash the green pods and blanche in boiling water for a minute and a half. Dunk them immediately in ice water for another minute and a half and store them in the freezer. Let the pods dry a little longer, and then walk on them to crush the pods. Winnow away the shells and store the seeds in airtight containers. The flowers can also be eaten raw or candied.

Parsnip

Uses: edible

Species: Parsnip (*Pastinaca sativa*)

Growing: Parsnips are a cold-climate crop, which need frost to fully develop. They can grow in any type of soil except the rockiest. The root is the edible part, and be aware that the leaves irritate the skin. Handling the plant during harvest requires gloves to prevent burning and blistering. The root is even more nutritious than carrots, and can be eaten raw or cooked. It is commonly boiled, roasted, or used in soups. The roots will be ready to harvest in mid-fall and taste much better after the first frost. Dig up the root and store in cold storage.

Passion Fruit

Uses: edible fruit, climbing vine

Species: Passion Fruit (*Passiflora edulis*)

Growing: There are two types of passion fruit: the yellow variety and the purple variety. The yellow kind is very big and grows fruit up to the size of a grapefruit, and the purple kind often tastes better but stays smaller than a lemon. They don't tolerate frost but are very cold hardy and can withstand almost freezing temperatures. Even if it does frost, a mature plant is so bushy that it insulates itself and recovers. It needs a strong trellis to support the fast-growing vine. The plants enjoy partial shade to full sun and well-drained fertile soil. Snails and a variety of diseases plague them, but with proper prevention a good crop is still possible. The fruit will turn ripe over a period of only a few days and will quickly turn purple or yellow and fall off. Gather the fruit from the tree and ground, wash and dry them without injuring them, and store them in cold storage. They will last up to 3 weeks. Freeze or juice the fruit to preserve it.

Paulownia

Uses: wood and timber, edible, bare soil erosion, pest control, bee forage

Species: Paulownia (*Paulownia fortunei*)

Growing: The paulownia is a very fast growing hardwood tree that can provide timber in as little as 5 years. It is touted as a solution to the world's unsustainable tropical wood industry. The leaves make excellent animal forage with high protein content, and they can be intercropped easily with grains and other plants. The plants need full sun and deep regular watering when the soil is dry. Additional fertilizer can be added throughout the year. In the first year the tree should grow to at least 10 feet (3 meters) tall. If it does not, cut it down in early spring before the leaves are formed. This will make it grow twice as fast the next year. The tree can be cut back like this almost every year. The

trees are only really susceptible to frost when young, but otherwise they can be grown in most climates, except for the very coldest places.

Pawpaw

Uses: edible fruit

Species: Pawpaw (*Asimina triloba*)

Growing: Some people inaccurately call this plant a papaya, but the pawpaw is not the same as the tropical fruit and is native to North America. It enjoys partial shade to full sun, and it is a pest resistant and generally hardy plant. It requires warm, humid temperatures, and provides large delicious fruit. Because of the ease of growing and its prolific supply of fruit, it is highly recommended for the home orchard in the right climate. It grows 20–35 feet (6–10 meters) tall and 20–35 feet wide and needs fertile, well-drained soil. Be aware that the seeds, leaves, and wood are toxic and sometimes made into pesticides. To harvest, wait until the fruit is just a little bit soft and has changed from green to yellow or brown. The fruit will keep for about 3 weeks in the fridge. It can be eaten raw or made into pie, bread, pudding, and jam, and can be used in place of bananas.

Pea

Uses: nitrogen fixing legume, edible, climbing vines

Species: Butterfly Pea (*Clitoria mariana*), Beach Pea (*Lathyrus japonicas maritime*), Everlasting Pea (*Lathyrus latifolius*), Earthnut Pea (*Lathyrus tuberosus*), Carolina Bush Pea (*Thermopsis villosa*)

Growing: Peas are easy to grow. They enjoy full sun and well-drained fertile soil, and they need trellising to keep them up. The peas will be ready to harvest in late summer and can be picked continuously until frost. The more you pick, the more will grow. The leaves of some of these peas are toxic. Some, like Snow Pea, are eaten fresh with the entire pod. Others are picked and shelled, and the peas are dried.

Peach

Uses: bee forage, edible

Species: Peach (*Prunus persica*), Nectarine (*Prunus persica var. nectarine*)

Growing: Peaches and nectarines need cold winters and long hot summers. At the same time, the spring flowers can't tolerate frost, preventing a fruit crop. This makes them an excellent temperate climate crop. They enjoy full sun, good airflow, regular watering, and frequent nitrogen fertilizer. When the fruit grows to almost an inch wide, it is common to thin it out so that the rest of the fruit will be sweeter. Pick the fruits on a hot day when they are just a little soft and have a sweet smell. They must be eaten or preserved immediately as they can't be refrigerated or stored in live storage.

Peanut

Uses: nitrogen fixing legume, edible

Species: Peanut (*Arachis hypogaea*)

Growing: Peanuts grow best in sandy soil and need a long growing season and regular watering. The pods must be harvested at just the right time, when they are fully ripe but before they detach into the ground. When this happens, the leaves will begin to turn yellow, and the peanut skin inside the pod will be like paper and of a light pink color. To harvest, dig up the entire plant gently and shake off the dirt. Then flip it upside down in a dry, shady place for a few days until the pods are mostly dry. Thresh and remove the pods from the plant. Peanuts may be eaten raw or roasted, ground into butter, or the oil extracted for use in cooking. Unshelled nuts will last for 9 months in the fridge, and in the freezer will last for years if blanched for 3 minutes and cooled in ice water for 3 minutes. To make peanut butter, roast the nuts for 20 minutes at 300°F (149°C), stirring now and then, and remove the shells. Process in a blender with half a teaspoon of salt per cup of peanut butter.

Pear

Uses: bee forage, edible fruit

Species: Asian Pear (*Pyrus bretschneideris*), European Pear (*Pyrus communis*)

Growing: Pears enjoy full sun. The Asian variety grows quite a bit taller than the European variety, but neither gets wider than 25 feet (8 meters). Pears are much hardier and pest-resistant than apples, with the exception of fireblight, which you can prevent by getting a resistant variety. Pears come in dwarf varieties, and there is a plethora of species to choose from. You will need 2 trees for pollination, but they can be from either species. Pears are somewhat unique in that you harvest them before they are ripe, with the only clue being that the stem will begin to detach. The pear will start to hang at an angle rather than vertically. The one thing apples have on them is that pears aren't as easy to store and preserve, although they are stored similarly. They need to be chilled, usually in a cold storage, in order to finish ripening, where they will stay good for a few months. They shouldn't be stored with onions, cabbage, carrots, or potatoes, or they will absorb the smells. They can also be made into jams and sauces.

Pecan

(see Hickory)

Persimmon

Uses: edible fruit

Species: American Persimmon (*Diospyros virginiana*)

Growing: The persimmon is the apple of Asia and is one of the most popular fruits. In North America it hasn't caught on, possibly because it tastes awful right until it reaches ripeness, at which point it becomes sweet and delicious. It prefers warmer, humid climates, and enjoys full sun. The size of the tree varies, and for pollination you will need a 1 male persimmon for every 8 female trees. There are hybrid varieties that grow larger fruit and come in dwarf sizes that may lend themselves to urban settings.

Pigeon Pea

Uses: wood and timber, edible, windbreak, nitrogen fixing legume, dryland tolerant

Species: Pigeon Pea (*Cajanus cajan*)

Growing: Pigeon peas are a very ancient source of food and grown all over the world. They can grow in most soil types and are resistant to drought, making them extremely hardy and versatile. They can be harvested as a pea and eaten raw, dried, or ground into flour. Their nutritional value is very high,

and when mixed with a cereal crop, peas make a very balanced source of protein. They are especially healthy when sprouted and cooked. They are also a very useful and easy to grow animal forage.

Plum

Uses: bee forage, windbreak, edible fruit

Species: American Plum (*Prunus Americana*), Canada Plum (*Prunus Americana* var. *nigra*), Chickasaw Plum (*Prunus angustifolia*), Hog Plum (*Prunus hortulana*), Beach Plum (*Prunus maritima*), Wild Goose Plum (*Prunus munsoniana*)

Growing: Plums enjoy full sun, and there is usually a species for most climates. They are susceptible to pests and diseases, and you need several plants for pollination. A small stand of plum trees can be very tasty if the right species is chosen, and hopefully without too much care to protect them. To harvest the plums, usually the first sign of ripening is a change in color, although that is still not a hard indication of readiness. They will also feel slightly soft, and the skin will acquire a powdery texture. The plums must be eaten immediately or processed into jam right away.

Pomegranate

Uses: dryland tolerant

Species: Pomegranate (*Punica granatum*)

Growing: There are many different varieties of pomegranate, but they are all very similar in the way that they grow. The differences lie mainly in the intended use and color of the fruit. They are very drought resistant and can tolerate freezing, although they need a long growing season to produce fruit. They enjoy partial shade to full sun. The fruit is ready to be harvested when it makes a metallic sound when tapped. If it isn't picked, then it will crack open and be ruined. They can't be picked, but rather must be clipped off close the fruit without leaving a stem. The fruits can be stored like apples in cold storage for 7 months as long as the temperature remains steady and the humidity does not rise too high. These can be eaten as they are by cutting them open and removing the juice sacs, or you can simply squeeze and juice them.

Poplar

Uses: streambank erosion control, wetland tolerant, wood and timber

Species: Eastern Cottonwood (*Populus deltoids*), Fremont Cottonwood (*Populus fremonti*), Black Poplar (*Populus nigra*), White Poplar (*Populus alba*), Narrowleaf Cottonwood (*Populus angustifolia*)

Growing: Poplars and cottonwoods grow in most climates and even in subarctic conditions, and they grow very quickly. Be careful not to plant them near any underground wires or pipes, as their extensive root system spreads up to 150 feet (46 meters) away from the tree. They enjoy fertile soil and regular watering, as well as partial shade to full sun. They are generally very simple to grow.

Potato

Uses: edible roots

Species: Potato (*Solanum tuberosum*)

Growing: Potatoes ripen in cool weather and stop when the temperature gets too hot, meaning they do better in northern areas. However, while they can withstand a light frost, they must be harvested before a heavy frost. They are planted from the *eyes*, or the indented brown spots, of other potatoes. During growth, the tubers tend to slowly pop out of the ground, but exposure to sunlight creates a green toxic spot that is inedible even with cooking. These potatoes must be covered up with more soil. Once they ripen, the whole plant is dug up carefully and cured. Curing requires keeping them at 65°F (18°C) for 10 days in a well-ventilated crate in a place that is very humid. Once cured, potatoes can be put in live storage or dried. Be aware that the rest of the plant is highly toxic, including any fruit that it might rarely produce above ground. The potato is a member of the nightshade family, and although it looks like a tomato plant, it is deadly. Potatoes can only be eaten cooked, and their green parts must not be eaten.

Prickly Pear

Uses: edible, dryland tolerant

Species: Prickly Pear (*Opuntia ficus-indica*), Eastern Prickly Pear (*Opuntia compressa*)

Growing: Prickly pears grow prolifically in the Sonoran desert and produce a beautiful edible fruit. Although it is a cactus, the prickly pear is very cold hardy and can grow in northern areas as well. It enjoys full sun and well-drained sandy soil. If it gets too much water, it will rot and collapse, so drainage is key. The entire plant is actually edible, as long as the cactus spines and seeds are removed. To harvest the *nopales*, or leaves, remove

▲ **Prickly pear fruit.**

one of the pads when they are about the size of your hand. This is done in spring and summer, and you must wear gloves. Pick off the spines and use a vegetable peeler or paring knife to remove the skin and eyes. This must be done carefully and thoroughly. Prickly pears are highly nutritious and can be eaten raw or cooked and used like green beans or a topping on Mexican dishes. To harvest the fruit, pick the smooth, firm, and shiny fruits, which will ripen for about one week in late fall or early winter. Wear gloves to do this. Use pliers to remove the spines, cut off the ends and use a paring knife to remove all of the skin layers from top to bottom. Be very careful to remove every part of the skin as it has tiny hairs that will hurt your insides. Remove the seeds. These can be eaten raw or cooked, or made into jam or juice.

Quinoa

Uses: edible seeds, dryland tolerant
Species: Quinoa (*Chenopodium quinoa*)
Growing: Quinoa is a hardy, protein-packed food source that can be grown in cooler climates. Not only are the leaves some of the most nutritious greens, but the seeds also make an excellent grain. Quinoa enjoys well-drained soil, is dryland tolerant, and prefers climates that don't get warmer than 90°F (32°C). You don't need to water it until it has 2 or 3 leaves and can grow with very little water. You can harvest some of the leaf greens when they are young, for salads. To harvest the seeds, wait until the leaves have fallen or even until right after the first frost. The only requirement is that the seeds must be very dry. If it rains, the seeds could germinate, so the timing must

be right. You should barely be able to make a dent with your thumbnail in the seed. Place the quinoa on a screen, rub to remove any debris, and then blow away the chaff with a fan. Quinoa must also be rinsed to remove the *saponin*, which is a bitter substance. Put the quinoa in a blender at the lowest speed and blend until it gets too soapy. Keep changing the water until it no longer gets frothy. Alternatively, you could put it in a pillowcase and run it in the cold water cycle of your washing machine. Cook the same as you would rice.

Radish

Uses: edible roots and shoots
Species: Radish (*Raphanus sativus*)
Growing: Radishes enjoy full sun and well-drained soil. There are many varieties, and they can be grown in most places. Some ripen in a month and can be replanted for several harvests throughout the season. Their long taproot breaks up hard soil, although it stunts their growth. Some radishes are harvested when small, and some grow to be the size of a potato. The entire plant is edible, and some are harvested for their spicy seeds and the greens used in salads.

Raspberry

Uses: bee forage, edible fruit
Species: Red Raspberry (*Rubus idaeus*), Black Raspberry (*Rubus occidentalis*), Creeping Raspberry (*Rubus tricolor*)
Growing: Everyone has eaten a raspberry, because there are hundreds of species everywhere. They self-pollinate and make good forest garden companions because they are less likely to take over than a blackberry. Trellising helps

to keep them under control, unless you want to use Creeping Raspberry as an effective groundcover. They enjoy full sun and well-drained, fertile soil. They are easy to grow and are considered a little invasive because they spread so quickly. They are a little susceptible to fungus and shouldn't be planted where tomatoes, potatoes, peppers, eggplants, or bulbs have been grown before. The berries are harvested when they change to their deepest hue and can be easily pulled from the branch. They come in gold, purple, black, or red so be aware of the species you have. These berries are eaten raw, frozen, or made into jam. The leaves are an effective herbal medicine. Raspberries are extremely nutritious and high in fiber.

Rhubarb

Uses: edible stalks, pest control
Species: Himalayan Rhubarb (*Rheum austral*), Rhubarb (*Rheum x cultorum*), Turkey Rhubarb (*Rheum palatum*)
Growing: Most gardeners grow rhubarb, which often persistently comes up in the compost pile. The leaves and roots are poisonous, but the stalks are delicious in pie (especially with strawberries). Rhubarb needs full sun and rich soil. The stalks can be harvested throughout the season. Only choose the firm, medium-sized stalks. Old stalks are too firm for cooking. Cut off the leaves and roots and discard in the compost pile.

Rosemary

Uses: bee forage, dryland tolerant
Species: Rosemary (*Rosmarinus officinalis*)

The Ultimate Guide to Permaculture

Growing: Rosemary is very easy to grow and very pest and drought hardy. It needs full sun and well-drained soil. It has often been used in making topiaries. The leaves can be harvested throughout the season as a culinary herb that is high in iron and calcium. It is also a popular medicinal herb, although it is toxic in high doses.

Russian Olive

Uses: edible fruit, nitrogen fixing legume, windbreak

Species: Russian Olive (*Elaeagnus angustifolia*)

Growing: Russian olive isn't related to olives, although its fruits look a little bit like them. It is considered an invasive species in many places because it spreads so quickly and does well even in some of the worst soils. It produces fruit in as little as 3 years. The fruit is edible although not particularly remarkable.

Sage

Uses: bee forage

Species: Common Sage (*Salvia officinalis*)

Growing: Sage is a common, easy to grow perennial herb that is used for its leaves in cooking and herbal medicine. The essential oil is commonly extracted for medicinal purposes. Sage enjoys partial shade to full sun and well-drained soil. It takes a year for it to get established, but it will continue to provide leaves for at least a few years.

Salsify

Uses: edible roots and shoots

Species: Salsify (*Tragopogon pornifolius*)

Growing: Salsify was once a more popular plant for vegetable gardens in North America, and it produces an edible root that tastes like oysters. The plant needs regular watering and loose soil. It is dug up in the fall to be eaten, or you can leave it in the ground over the winter to let it sprout again in the spring. The root must be peeled and washed and put in water with lemon juice or vinegar to prevent oxidation. Then it can be boiled and mashed, fried, steamed, or used in soups or stew.

Sesbania

Uses: wood and timber, edible, windbreak, nitrogen fixing

Species: Rattlebox (*Sesbania drummondi*), Spiny Sesbania (*Sesbania bispinosa*), Egyptian Pea (*Sesbania sesban*), Rostate Sesbania (Sesbania rostata)

Growing: Sesbania is a large shrub that is a common green manure and animal forage crop around the world, and it is high in protein. It can tolerate terrible soil, flooding, and short periods of drought. It is hardy to light frost, but heavy frost will kill it. To store for animal fodder, cut it when it is at least 3 feet (1 meter) tall and process in the same way as hay. It may not be the best field forage crop for goats, which can kill it.

Siberian Pea Shrub

Uses: seedpod and sugar source, windbreak, nitrogen fixing legume, bee forage

Species: Siberian Pea Shrub (*Caragana arborescens*)

Growing: This plant is a large shrub or small tree that grows in most areas and is tolerant to extreme cold. It reaches up to 16 feet (5 meters) tall. The pods have edible beans that are a little too much trouble for people but make excellent chicken fodder. If you do want to eat them, the pods must be cooked, or the flowers can be added to salads. The entire plant is used in herbal medicine.

Sloe

Uses: bee forage, edible, wood and timber
Species: Sloe (*Prunus spinosa*)
Growing: Also known as blackthorn, sloe is a large shrub that grows up to 15 feet (5 meters) tall and produces purple fruit. Sloes look much like cherry plums, but the flowers of sloe are cream rather than white and bloom later. Sloe has large thorns, which make it a great hedge plant, and the wood is useful for firewood or carpentry projects. The fruit can be used in juice, wine, or jam, and some recommend that these should be harvested after the first frost for better flavor.

Sorghum

Uses: edible, bee forage, dryland tolerant
Species: Sorghum (*Sorghum bicolor*)
Growing: Sorghum is very hardy and tolerates drought and high temperatures. It can be used as a grain, processed into molasses, and used as animal fodder. It needs a long growing season with warm weather and lots of fertilizer. As a grain it is raised and harvested the same as wheat or corn, although you might need to wait until the first frost to allow it to dry enough. The plants cannot be used as animal fodder until they are at least 18 inches (0.5 meters) tall as the young shoots are poisonous. To make your own molasses, chop them down at the ground and remove the leaves. The stalks, or *canes*, must be pressed or ground, and the juice is caught

▾ Sloe berries

The Ultimate Guide to Permaculture

in a container. This juice is then boiled down like maple syrup until it is highly concentrated and sweet. This must be stirred continuously to prevent scorching. The syrup can be stored for many months, although it will begin to harden at the bottom. This hard sugar can be used as a sweetener or candy as well.

Stone Pine

Uses: edible nut, dryland tolerant
Species: Stone Pine (*Pinus pinea*)
Growing: The stone pine can grow 60 feet (18 meters) tall or more and produces one of the most ancient foods: the pine nut. It is a beautiful tree with a characteristic umbrella growing high above the tall stalk. Stone pines are very hardy, tolerant to drought, and take a while to grow. They enjoy full sun and well-drained soil. When a pine does finally produce cones, the seeds are edible and often used in Italian pasta sauces.

Strawberry

Uses: edible fruit, ground cover
Species: Garden Strawberry (*Fragaria x ananassa*), Beach Strawberry (*Fragaria chiloensis*), Musk Strawberry (*Fragaria moschata*), Alpine Strawberry (*Fragaria vesca alpine*), Wild Strawberry (*Fragaria virginiana*)
Growing: Strawberries generally need full sun and well-drained fertile soil in a temperate climate. They produce some of the tastiest fruit in the world. They are susceptible to pests and weather. Alpine does not spread like the other species do. Garden strawberries are the most popular, but they must be rotated every three years

to stop diseases and fungus. Beaches and Musks work well as an effective groundcover.

Sumac

Uses: edible fruit
Species: Staghorn Sumac, (*Rhus* typhina), Fragrant Sumac (*Rhus aromatica*), Smooth Sumac (*Rhus glabra*)
Growing: Sumac enjoys partial shade to full sun with well-drained soil. It grows easily in any kind of soil. The berries are edible and are usually dried and then ground up to form a powder that is used to season rice or meat, and in some places they are used to make a drink like lemonade. The most effective way to contain sumac is with goats, as it tends to spread and will just grow more if mowed. Also, use caution as poison ivy, poison oak, and poison sumac are all part of the same family and have some similarities.

Sunflower

Uses: edible
Species: Sunflower (*Helianthus annuus*)
Growing: Sunflowers need full sun and fertile, well-drained soil with regular watering. The seeds are harvested at the end of summer from the dried flower heads and can be eaten raw or roasted, ground into butter, or the oil can be extracted for cooking. The leaves can be fed to cattle. Sunflowers are very easy to grow, as long as they get enough sun.

Sweetfern

Uses: ground cover, nitrogen fixing legume, dryland tolerant
Species: Sweetfern (*Comptonia peregrina*)

Growing: The sweetfern is not a fern but rather a shrub. It enjoys well-drained soil and is hardy to cold. It enjoys partial shade to full sun. The leaves are used as an herb in cooking or for medicinal purposes. It also produces an edible fruit that may be eaten raw or cooked.

Sweet Woodruff

Uses: wetland tolerant, edible

Species: Sweet Woodruff (*Galium odoratum*)

Growing: Sweet woodruff is an herb that enjoys rich, moist soil and regular watering. It makes an excellent garden border as deer won't eat it. It has a strong scent that is used in potpourri and repels moths. It is edible but not in high doses. It is commonly used as an herbal medicine.

Tagasaste

Uses: bee forage, windbreak, dryland tolerant, nitrogen fixing legume, edible

Species: Tagasaste (*Chamaecytisus palmensis*)

Growing: Tagasaste is a small evergreen tree that enjoys well-drained soil. It is hardy to cold and hot temperatures, although frost will damage it or kill young seedlings. It is a very valuable animal forage crop. Sheep and cattle can graze the plants without killing them and without irrigation for many years. It has a similar nutritional value to alfalfa, and is full of protein and other minerals.

Taro

Uses: edible roots and shoots, wetland tolerant

Species: Taro (*Colocasia esculenta*)

Growing: Taro can be grown in a paddy, like rice, to keep control of the weeds, but the water must be cool and flowing. However, a marshy place also works. Taro also enjoys a long, hot summer. To harvest, wait until the leaves turn yellow and dig them up. Make sure that you are eating the edible varieties and not the ornamental ones. The plant must be well cooked, as the raw parts can cause a burning sensation in your mouth and throat. To prepare them, soak overnight in cold water and then boil, bake, or roast like potatoes. In the United States the taro is sometimes known as Dasheen and cooked, dried,

▼ Taro

and ground into flour. It can also be stored in live storage like potatoes. The leaves are also edible and often cooked like kale or used in traditional Hawaiian dishes.

Thyme

Uses: ground cover, edible, bee forage
Species: Woolly Thyme (*Thymus psuedolanuginosus*), Citrus Thyme (*Thymus citriodorus*), Wild Thyme (*Thymus serpyllum*), Common Thyme (*Thymus vulgaris*)
Growing: Thyme is a perennial herb that enjoys full sun and well-drained soil. It tolerates drought and cold northern temperatures. It is used in cooking and

herbal medicine and may be used raw or cooked. Woolly thyme makes the densest ground cover. The leaves may be harvested throughout the season, as they will grow back quickly.

Turnip

Uses: edible roots and shoots
Species: Turnip (*Brassica rapa*)
Growing: The entire turnip plant is edible, with the root being eaten like potatoes, and the greens being cooked like kale. Turnips prefer well-drained soil, but they are extremely versatile and are grown in just about every climate and condition. Turnips can also serve as animal fodder. They

are harvested at the end of the season, left in the ground over the winter, or put in live storage.

Vetch

Uses: bare soil erosion control, nitrogen fixing legume, ground cover

Species: Hairy Vetch (*Vicia villosa*), Bitter Vetch (*Lathyrus linifolius montanus*), American Vetch (*Vicia americana*), Wood Vetch (*Vicia caroliniana*), Tufted Vetch (*Vicia cracca*), Sweet Vetch (*Hedysarum boreale*), Milk Vetch (*Atragalus glycyphyllos*)

Growing: As vetch is often grown in conjunction with clover and alfalfa as a ground cover, it sometimes becomes part of a pasture that animals forage in. However, vetch is toxic over periods of time. If your animals eat too much of it for too long, they can end up with a nervous system disorder. Despite this, it is still useful because it is the most cold hardy cover crop.

Violets

Uses: ground cover, edible leaves

Species: Canada Violet (*Viola Canadensis*), Labrador Violet (*Viola labradorica*), Sweet Violet (*Viola odorata*)

Growing: These cute flowers have edible leaves, but not all are created equal. They grow just about anywhere, but local native species will vary in flavor. They act as a hardy evergreen ground cover. Sweet violets are the sweetest and do the best where it is not too hot.

Walnut

Uses: edible nut, wood and timber

Species: Heartnut or Japanese Walnut (*Juglans ailantifolia* var. *cordifolia*), Butternut or White Walnut (*Juglans cinerea*), Black Walnut (*Juglans nigra*)

Growing: Walnuts enjoy full sun and grow 60–100 feet (18–30 meters) tall and 50–100 feet (15–30 meters) wide. Black walnut improves the soil by building mineral content, unlike the other varieties, but they all taste delicious. Black walnut grows from the Midwest to the East coast, extending into the southern United States, the Butternut grows in the Northeast, radiating out from the Great Lakes, and the Heartnut grows in the coldest northern areas. Black walnut is considered the most valuable because it is the scarcest. Walnuts need fertile, moist soil. To harvest, wait until the walnuts begin turning from green to brown and start falling off the tree. The ones that are ready are partly brown and partly green. If they are completely black, they may not be usable. These must then be hulled by crushing them. Some people do this with their car, by stomping on them, or by using converted cement mixers for this purpose. Once hulled, they must be dried for 2 months in a place with good air circulation. Place them in burlap or other breathable bag loosely. After they are dry, they can then be cracked like any regular nut.

Water Chestnut

Uses: edible, water plant

Species: Water Chestnut (*Eleocharis dulcis*)

Growing: The water chestnut is not a nut but rather a water plant that grows an edible vegetable that is small and rounded, looking somewhat like a nut. Water chestnuts can be eaten raw, boiled, cooked, or pickled. They enjoy a long growing season and full sun, and they should be planted around the shallow edges of a pond, with the water level staying at least 2 inches above the soil. At the end of summer the leaves will die, and you simply dig up the plant. Rinse them off and use them accordingly.

Watercress

Uses: streambank erosion control, wetland tolerant, water plants

Species: Watercress (*Nasturtium nasturtium officinale*)

Growing: Although the Latin name is *Nasturtium*, watercress is not closely related to Nasturtium flowers. This highly nutritious plant can be grown in wetlands or ponds, as long as the soil is fertile and water is abundant. It enjoys partial shade and warm temperatures. Harvest the leaves throughout the season and use fresh.

Wild Ginger

Uses: wetland tolerant, edible roots and shoots, ground cover, edible root

Species: Wild Ginger (*Asarum canadense*), Shuttleworth's Wild Ginger (*Asarum shuttleworthii*)

Growing: A lovely groundcover that grows natively in most areas but still doesn't quite have the same flavor as cultivated varieties. Only the root is edible; the leaves are toxic. It is actually not related to ginger root but is called ginger because it tastes and smells like ginger. It prefers moist, fertile soil and shade. It is hardy in cold climates.

Wild Rice

Uses: water plants

Species: Wild Rice (*Zizania aquatica*), Northern Wild Rice (*Zizania palustris*)

Growing: Rice growing has been described elsewhere in this book as a grain crop, but it can also be grown as an incidental water plant. Wild rice is actually not closely related to the common Asian rice (*Oryza sativa*). It is very nutritious. The seeds are simply thrown into a pond and sink to the bottom to sprout, and wild rice enjoys full sun and cooler climates. It does not compete well with cattails and should be kept separate. The rice is delectable to birds and other foragers, so it may be prudent to protect it. Commercial growers drain their paddies to harvest the rice, but that's not an option for the small pond owner. Native people would use a canoe, but if you can't do that, just make sure the rice is planted within arm's reach of the shore. Then in the fall when the seed heads have fully developed and fall off easily, use a stick to tap them into a basket. These must then be laid out to dry in the sun for at least a day and then parched. This is traditionally done in a cast iron pot over a wood fire, and the seeds must be stirred continuously until they cahnge from green to a darker color. This takes a couple of hours. Use just like you would other types of rice, but add 3 cups of water for every cup of rice.

Willow

Uses: streambank erosion control, wetland tolerant, water plants, bee forage, wood and timber

Species: Willow (*Salix L.*)

Growing: Willows are beautiful trees with many uses. The bark contains salicylic acid and is used along with the leaves n herbal medicine. It has tough, flexible wood used in making a variety of items. The plant grows easily from cuttings or from broken branches that fall on the ground, and it grows in most climates. Willows prefer moist soil and are often planted on stream banks to stop erosion with their many tangled roots. In an urban area these roots can cause trouble by destroying pipes and wires, so use caution. There are hundreds of species, so use the variety that is native to your area.

Yarrow

Uses: bee forage, ground cover, edible
Species: Yarrow (*Achillea millefolium*)
Growing: Yarrow enjoys partial shade to full sun and provides beautiful ground cover that gives a home to beneficial insects. It has many herbal medicinal uses and benefits the soil. However, be aware that too much yarrow intake can cause some severe digestive problems, so take in moderation.

▲ Yarrow

Soil-Improving Plants

	Legumes	Nitrogen fixers	Bare soil erosion	Stream bank erosion	Wetland tolerant	Water plants	Dryland tolerant	Ground cover	Edible
Acacia	X	X	X				X		X
Alder	X	X							
Alfalfa	X	X							X
Almond							X		X
Amaranth							X		X
American yellowwood	X	X							
Asparagus				X					X
Autumn olive	X	X	X						X
Bamboo	X		X	X					X
Beans	X	X							X
Bearberry								X	X
Bishop's weed								X	X
Black locust	X	X	X				X		
Black tupelo			X	X	X				
Bunchberry	X			X	X			X	X
Carob	X	X					X		X
Cattail						X			X
Ceanothus	X	X							X
Clover	X	X						X	X
Cranberry						X			X
Daylily			X		X				X
Duck potato						X			X

	Legumes	Nitrogen fixers	Bare soil erosion	Stream bank erosion	Wetland tolerant	Water plants	Dryland tolerant	Ground cover	Edible
Duckweed						x			
Fenugreek	x	x							x
Fig							x		x
Flowering dogwood					x			x	x
Gliricidia	x	x							x
Groundnut	x	x							x
Hawthorn			x				x		x
Hairy vetch	x	x	x						
Holm oak							x		
Honey locust		x					x		x
Jasmine					x				x
Japanese pagoda tree	x	x							x
Jujube							x		x
Lavender							x		
Leucaena	x	x		x				x	x
Lemongrass				x					x
Licorice	x	x							x
Lotus						x			x
Lowbush blueberry								x	x
Lupine	x	x						x	x
Mesquite	x	x					x		x

	Legumes	Nitrogen fixers	Bare soil erosion	Stream bank erosion	Wetland tolerant	Water plants	Dryland tolerant	Ground cover	Edible
Mint					X			X	X
Mulberry							X		X
Nasturtium								X	X
Nettle								X	X
N.Z. spinach							X	X	X
Olive							X		X
Palo Verde	X	X	X				X		X
Paulownia			X						X
Pea	X	X							X
Peanut	X	X							X
Persian ground ivy								X	
Pigeon pea	X	X					X		X
Pomegranate							X		X
Poplar				X	X				X
Prickly pear							X		X
Quinoa							X		X
Raspberry								X	X
Rice						X			X
Rosemary							X		X
Rush						X			
Russian olive	X	X	X				X		X
Siberian pea shrub	X	X							X

	Legumes	Nitrogen fixers	Bare soil erosion	Stream bank erosion	Wetland tolerant	Water plants	Dryland tolerant	Ground cover	Edible
Sesbania		X			X		X		X
Sorghum							X		X
Stone pine							X		X
Strawberry								X	X
Sunflower									X
Sweetfern	X	X							X
Sweet potato							X	X	X
Sweet woodruff					X				X
Taro					X				
Tagasaste	X	X					X		X
Thyme								X	X
Vetch	X	X							X
Violet								X	X
Watercress				X	X	X			X
Wild ginger					X			X	X
Wild rice						X			X
Willow				X	X	X			X
Yarrow								X	X

Other Uses

	Bee forage	Pest control	Wood/Timber	Windbreak	Climbing vines	Edible
Acacia			X			X
Alfalfa	X					X
Almond	X					X
Amaranth						X
American linden	X		X			X
Apple	X		X			X
Apricot	X					X
Autumn olive	X			X		X
Bamboo			X			X
Bean	X				X	X
Bee balm	X					
Beet						X
Borage	X					X
Black locust	X		X	X		
Black tupelo	X		X			
Blueberry	X					X
Buckwheat	X					X
Carrot						X
Cedar		X	X			
Cherry	X		X			X
Chestnut			X			X
Chicory	X					X
Clover	X					X
Comfrey	X					

	Bee forage	Pest control	Wood/Timber	Windbreak	Climbing vines	Edible
Cranberry	X					X
Currant						X
Daisy		X				
Dandelion	X					X
Daylily	X					X
Dill		X				X
Fennel		X				X
Flowering dogwood			X			X
Gliricidia	X		X			
Gooseberry	X					X
Grape					X	X
Groundnut					X	X
Guava						X
Hawthorn	X			X		
Hazelnut						X
Hickory			X			X
Honeysuckle	X				X	
Hops						X
Horseradish						X
Hyssop	X					X
Jasmine					X	X
Jerusalem artichoke						X
Jicama						X
Jujube			X			X
Kiwi fruit					X	X

	Bee forage	Pest control	Wood/Timber	Windbreak	Climbing vines	Edible
Lavender	X	X				
Leucaena			X			
Loganberry	X					X
Loquat						X
Lupine	X					
Macadamia						X
Maple	X		X			X
Marigold		X				
May apple						X
Mericrest						
Mesquite	X					
Mint	X	X				X
Moringa						X
Myrtle						
Natal plum						X
Neem		X				
Oak			X			X
Palm			X			X
Parsnip						X
Passion fruit					X	X
Paulownia		X	X			X
Pawpawz		X				X
Peach	X					X
Pear	X					X
Pigeon pea			X	X		X

	Bee forage	Pest control	Wood/Timber	Windbreak	Climbing vines	Edible
Plum	x			x		x
Poplar			x	x		
Potato						x
Quinoa						x
Rhubarb		x				x
Rosemary	x	x				x
Russian olive				x		x
Sage	x					x
Salsify						x
Sesbania			x	x		x
Siberian pea shrub	x			x		x
Sloe	x					
Sorghum	x					x
Sumac						x
Taro						x
Tagasaste	x			x		x
Tea Tree		x				
Turnip						x
Vetch					x	
Walnut			x			x
Water chestnut						x
Willow	x					
Yarrow	x					x

Bibliography

BlueScope Water. (n.d.) Rain Water Tanks—Life Cycle Analysis. *BlueScope Steel Australia.* Retrieved from www.bluescopesteel.com.au/building-products/rainwater-harvesting/life-cycle-analysis-for-rainwater-tanks

Bohnsack, Ute. (n.d.) Companion Planting Guide. *South East Essex Organic Gardeners.* Retrieved from www.gb0063551.pwp.blueyonder.co.uk/seeog/companion/

British Columbia Ministry of Public Safety and Solicitor General. (n.d.) Flood Proofing Your Home: Minimize Damage if a Flood Strikes Your Family Home. *Emergency ManagementBC.* Retrieved from www.pep.bc.ca/hazard_preparedness/flood_tips/Floodproof.pdf

Burns, Russel and Barbara Honkala. (1990) Black Walnut. *Silvics of North America.* U.S. Department of Agriculture, Forest Service. Retrieved from www.na.fs.fed.us/pubs/silvics_manual/volume_2/juglans/nigra.htm

Clarke, S. (2003) Electricity Generation Using Small Wind Turbines at Your Home or Farm. *Engineering Factsheet.* Ontario Ministry of Agriculture, Food and Rural Affairs. Retrieved from www.omafra.gov.on.ca/english/engineer/facts/03-047.htm

Cook, Howard. (2011) *The Principles and Practice of Effective Seismic Retrofitting.* Bay Area Retrofit. Retrieved from www.bayarearetrofit.com/PDFs/design_book.pdf

Cunningham, Sally Jean. (1998) *Great Garden Companions: A Companion Planting System for a Beautiful, Chemical-Free Vegetable Garden.* Rodale.

Denzer, Kiko. (2001) Build Your Own Wood-fired Earth Oven. *Mother Earth News.* Retrieved from www.motherearthnews.com/Do-It-Yourself/2002-10-01/Build-Your-Own-Wood-Fired-Earth-Oven.aspx?page=5

Deveau, Jean Louis. (2002) Raising Quail for Food in Fredericton, New Brunswick, Canada. *Urban Agriculture Notes.* Retrieved from www.cityfarmer.org/quail2.html

Down Garden Services. (2011) Companion Planting. *Down Garden Services.* Retrieved from www.dgsgardening.btinternet.co.uk/companion.htm

Exploratorium. (n.d.) Basic Sourdough Starter. *Science of Cooking.* Retrieved from www.exploratorium.edu/cooking/bread/recipe-sourdough.html

Faires, Nicole. (2011) *The Ultimate Guide to Homesteading: An Encyclopedia of Independent Living.* Skyhorse Publishing.

Forcefield. (2008) The Bottom Line About Wind Turbines. *Otherpower.com.* Retrieved from www.otherpower.com/bottom_line.shtml

Fowler, D. Brian. (2002) Winter Wheat Production Manual: Chapter 10— Growth Stages of Wheat. *University of Saskatchewan College of Agriculture and Bioresources.* Retrieved from www.usask.ca/agriculture/plantsci/winter_cereals/Winter_wheat/CHAPT10/cvchpt10.php

Fraas, Wyatt. (n.d.) Beginning Farmer and Rancher Opportunities. *Center for Rural Affairs.* Retrieved from www.cfra.org/resources/beginning_farmer

Fukuoka, Masanobu. (1985) *The Natural Way of Farming: The Theory and Practice of Green Philosophy.* Japan Publications.

Garden Action. (2010) Plum Tree Care. *GardenAction*. Retrieved from www.gardenaction.co.uk/fruit_veg_diary/fruit_veg_mini_project_march_2_plum.asp#plum_start

Gasparotto, Suzanne. (n.d.) Organically-Raised Goats. *Onion Creek Ranch*. Retrieved from www.tennesseemeatgoats.com/articles2/organicGoats06.html

Golden Harvest Organics. (2011) Companion Planting. *Golden Harvest*. Retrieved from www.ghorganics.com/page2.html

Hackelman, Michael and Claire Anderson. (2002) Harvest the Wind. *Mother Earth News*. Retrieved from www.motherearthnews.com/Renewable-Energy/2002-06-01/Harvest-the-Wind.aspx

Hart, Kelly. (n.d.) Frequently Asked Questions. *Earthbag Building*. Retrieved from www.earthbagbuilding.com/faqs.htm

Hemenway, Toby. (2009) *Gaia's Garden: A Guide to Home-Scale Permaculture* Chelsea Green.

Hill, Stuart B. (1975) Companion Plants. *Ecological Agriculture Projects*. Retrieved from http://eap.mcgill.ca/publications/EAP55.htm

Holmgren, David. (2009) *Permaculture: Principles & Pathways Beyond Sustainability*. Holmgren Design Services

Hooker, Will. "Introduction to Permaculture" North Carolina State University, Distance Education. 18 Mar. 2011

Jacke, Dave and Eric Toensmeier. (2005) *Edible Forest Gardens Volume 1: Vision & Theory*. Chelsea Green.

Jacke, Dave and Eric Toensmeier. (2005) *Edible Forest Gardens Volume 2: Ecological Design and Practice for Temperate Climate Permaculture*. Chelsea Green.

Jason, Dan. (2011) Growing Amaranth and Quinoa. *Salt Spring Seeds*. Retrieved from www.saltspringseeds.com/scoop/powerfood.htm

Jenkins, Joseph. (2005) *The Humanure Handbook: A Guide to Composting Human Manure*. Joseph Jenkins.

Kitsteiner, John. (2011) Permaculture Projects: Coppicing. *Temperate Climate Permaculture*. Retrieved from www.tcpermaculture.com/2011/06/permaculture-projects-coppicing.html

Luzadis, V. A. and E. R. Gossett. (1996). Sugar Maple. *Forest Tress of the Northeast*. Pages 157–166. Cornell Media Services.

Lynch, William. (2002). Fish Species Selection for Pond Stocking. *Ohio State University Extension Factsheet*. Retrieved from http://ohioline.osu.edu/a-fact/pdf/0010.pdf

Madison, Deborah. (1999) *Preserving Food Without Freezing or Canning: Traditional Techniques Using Salt, Oil, Sugar, Alcohol, Vinegar, Drying, Cold Storage and Lactic Fermentation*. Chelsea Green.

Mars, Ross. (2005) *The Basics of Permaculture Design*. Chelsea Green.

Permaculture. (2011) About the PDC Certificate. *Midwest Permaculture*. Retrieved from http://midwestpermaculture.com/about/certification/

Missouri Department of Conservation. (1999) Fathead Minnows in New Ponds and Lakes. *Aquaguide*. Retrieved from

http://mdc.mo.gov/sites/default/files/
resources/2010/05/4890_2843.pdf

Mollison, Bill. (1988) *Permaculture*. Tagari.

Mollison, Bill and Reny Slay. (2009)
Introduction to Permaculture. Tagari.

Mohamed Lahlou. (2000) Slow Sand
Filtration. *Tech Briefs*. Retrieved
from www.nesc.wvu.edu/pdf/DW/
publications/ontap/tech_brief/TB15_
SlowSand.pdf

New Brunswick Department of
Environment. (2011) Composting.
*Backyard Magic: The Composting
Handbook*. Retrieved from www.gnb.
ca/0009/0372/0003/index-e.asp

Nova Scotia Museum. (n.d.) Lupin
or Lupine (Lupinus Species).
The Poison Plant Patch. Retrieved
from http://museum.gov.ns.ca/
poison/?section=species&id=99

North Dakota State University. (2011)
Insulation. *Bioenvironmental
Engineering*. Retrieved from www.
ageng.ndsu.nodak.edu/envr/Insulatn.
htm

OK Solar. (2011) Angle of Orientation for
Solar Panels & Photovoltaic Modules.
OkSolar.com Technical Information.
Retrieved from www.oksolar.com/
technical/angle_orientation.html

Olsen, Ken. (2001) Solar Hot Water: A
Primer. *Arizona Solar Center*. Retrieved
from www.azsolarcenter.org/tech-
science/solar-for-consumers/solar-hot-
water/solar-hot-water-a-primer.html

Patterson, John. (n.d.) Solar Hot Water
Basics. *Homepower Magazine*.
Retrieved from http://homepower.com/
basics/hotwater/

Pennsylvania Fish and Boat Commission.
(n.d.) Using Mussels to "Clean" a Pond.
Q & A. Retrieved from www.fish.state.

pa.us/images/pages/qa/misc/mussels_
pond.htm

Permaculture Institute. (2008) *Permaculture
Design Certificate Course Outline*.
Retrieved from www.permaculture.org/
nm/images/uploads/PDC_cert_book_.
pdf

Permaship. (2011) Graywater in the
Garden. *Permaculture Projects*.
Retrieved from https://sites.google.
com/site/permaship1/permaculture-
practice/graywater-garden

Permaship. (2011) Permaculture Pond.
Permaculture Projects. Retrieved
from http://sites.google.com/site/
permaship1/permaculture-practice/
permaculture-pond

Pushard, Doug. (2010) Rainwater
Harvesting—Pumps or Pressure Tanks.
HarvestH2O. Retrieved from www.
harvesth2o.com/pumps_or_tanks.shtml

Rodale Press. (1977) *The Rodale Her Book:
How to Use, Grow, and Buy Nature's
Miracle Plants*. Rodale.

Rolex Awards. (2005) Ancient Technology
Preserves Food. *Rolex Awards*.
Retrieved from www.rolexawards.com/
en/the-laureates/mohammedbahabba-
the-project.jsp

Rombauer, Irma and Marion Becker. (1975)
The Joy of Cooking. Bobbs-Merril.

Roose, Debbie. (2011) Selling, Eggs, Meat,
and Poultry in North Carolina: What
Farmers Need to Know. *Growing Small
Farms*. Retrieved from: www.ces.ncsu.
edu/chatham/ag/SustAg/meatandeggs.
html

Royal Horticultural Society. (2011)
Plums: Pruning. *Gardening for all*.
Retrieved from http://apps.rhs.org.uk/
advicesearch/Profile.aspx?pid=339

Schafer, William. (2009) Making Jelly.
University of Minnesota Extension.

Retrieved from www.extension.umn. edu/distribution/nutrition/dj0686.html

Solar Cooking International. (2011) Solar Cooking Info. *Solar Cookers World Network*. Retrieved from http://solarcooking.wikia.com/wiki/Solar_Cookers_World_Network_(Home)

Spiers, Adrian. (1997) Pruning to Prevent Silverleaf. *The Horticultural and Food Research of New Zealand*. Retrieved from www.hortnet.co.nz/publications/hortfacts/hf205016.htm

Stark, Kevin. (n.d.) Use of Chinampa Agriculture by the Aztecs. *Anthropology 387: The Aztecs. Pacific Lutheran University*. Retrieved from www.plu.edu/~starkkl/anthropology-387/home.html

USDA, NRCS. (2011) The PLANTS Database. *National Plant Data Team*. Retrieved from http://plants.usda.gov

USDE. (2011) Natural Fiber Insulation Materials. *Energy Savers*. Retrieved from www.energysavers.gov/your_home/insulation_airsealing/index.cfm/mytopic=11560

USDE. (2011) Microhydropower Turbines, Pumps and Waterwheels. *Energy Basics*. Retrieved from www.eere.energy.gov/basics/renewable_energy/turbines_pumps_waterwheels.html

USGS. (2011) The Water Cycle. *USGS Water Science for Schools*. Retrieved from http://ga.water.usgs.gov/edu/watercyclesummary.html

Virginia Water Resources Research Center. (2011) Vegetated Emergency Spillway. *Virginia Stormwater BMP Clearinghouse*. Retrieved from http://vwrrc.vt.edu/swc/NonPBMPSpecsMarch11/Introduction_App%20C_Vegetated%20Emergency%20Spillways_SCraftonRev_03012011.pdf

Weinmann, Todd. (n.d.) Companion Planting. *Cass County Extension, North Dakota State University*. Retrieved from www.ag.ndsu.edu/hort/info/vegetables/companion.htm

Wheaton, Paul. (2011) Permaculture Articles. *Rich Soil*. Retrieved from www.richsoil.com/

White, Mel. (1976) Cure Your Own Olives. *Mother Earth News*. Retrieved from www.motherearthnews.com/Real-Food/1976-01-01/Cure-Your-Own-Olives.aspx

Yocum, David. (n.d.) Design Manual: Greywater Biofiltration Constructed Wetland System. *Bren School of Environmental Science and Management, University of California*. Retrieved from http://fiesta.bren.ucsb.edu/~chiapas2/Water%20Management_files/Greywater%20Wetlands-1.pdf

Index

AC, 47-48, 50

Acacia, 179, 193, 253

Acorn, 234, 283

Addictive behaviors, 7

Adobe house, 84, 88

Air drying, 151

Air insulation, 86

Alder, 105, 235, 253, 301

Alfalfa, 119, 191, 193, 199, 204, 213, 229-230, 253-254, 276, 296, 298, 301, 305

Almond, 254, 301, 305

Amaranth, 119, 124, 193, 254, 301, 305

American linden, 254-255, 305

American yellowwood, 255

Animal forage, 11, 245, 262, 275, 279, 287, 289, 293, 296

Anise, 119, 124, 140

Annuals, 107, 138, 282

Apple, 115, 118, 223, 128, 153, 160, 165, 185, 229, 255, 268, 278, 279, 305, 307

Appliances, 47, 82-83

Apricot, 137, 185, 255, 305

Aquaculture, 15, 57, 61, 66, 68-69, 247

tank, 15, 61

Aquatic nursery, 245

Arable land, 248-249

Arid climate, 28

Array, 47-48, 53

Artichoke, 115, 152, 158, 174, 272, 306

Artificial pollen, 182-183

Ash, 108, 235

Asparagus, 119, 129, 128-129, 134, 152, 255, 301

Autumn olive, 112, 193, 256, 301, 305

Avocado, 137, 256

Azolla, 256

Bacteria in food, 73, 148-149, 158-159, 161, 209, 273

Bad soil, 108-109

Balsa insulation, 86

Bamboo, 19, 31, 35, 66, 84-85, 93, 108, 152, 172, 174, 232, 235, 245, 256-257, 301, 305

Bamboo house, 35, 84-85

Barn, 72, 79, 169, 182, 191-192203, 206, 211, 215, 217-209, 224-225, 230

flooring, 218-219

Barrier dam, 63-64

Basil, 119, 123, 130, 134, 140, 276

Bass, 69-71

Batch hot water, 44

Bath, 78

Bathroom, 35, 78-79, 82, 90, 99

Battery, 47-48, 51, 53

Bean sprout, 152

Beans, 94, 115, 135, 137-138, 148, 153, 156, 160, 173, 193, 225-226, 257, 265, 273, 276, 291, 293, 301

Bearberry, 257-258, 301

Bed design, 169

Bee, 19, 26, 30, 115, 121, 125, 126, 132, 133, 170, 178-184

allergy, 179

bread, 183

calendar, 183-184

care, 40, 117, 178-184

equipment, 179-180

feeding, 40

forage, 18, 179

handling, 181

health, 183

hive, 21, 181, 183

hive splitting, 183

in permaculture, 178-179

Bee balm, 258

Beech, 235

Beet, 106, 119, 120, 122-124, 126-128, 130-132, 151-152, 158, 160, 173, 204, 235, 258, 305,

Beet greens, 152, 258

Beneficial plants, 13

Bentonite, 108

Berries, 13, 105, 106, 164, 179, 193, 198, 200, 227-229, 244, 245, 257-260, 263-264, 267, 269-270, 275, 280-282, 292, 294-295

Bill Mollison, xiii, 1, 3, 167,

Biogas, 46

Biomass, 30, 38, 43, 46-47, 79, 111

Birch, 235

Bishop's weed, 258-259, 301

Black eyed peas, 152

Black locust, 109, 235, 259, 269, 301, 305

Black tupelo, 259, 301, 305

Black walnut, 109, 130, 134, 298

Blackberry, 13, 105, 179, 200, 229, 259, 275, 282, 292

Blackwater, 99, 101

Blanching, 151-152

Blueberry, 227-228, 259, 264, 302, 305

Bluegill, 69-71

Borage, 121, 132-134, 260, 305

Botulism, 148, 158

Brassica, 118-128, 130, 132-134, 138, 297

Broad bean, 138, 152

Broadcast, 169, 221, 230

Broccoli, 118, 138, 140, 152, 173, 270

Brood cell, 181-183

Brood chamber, 180

Broodiness, 196

Brussels sprout, 119, 152, 158, 160, 173,

Buckwheat, 113, 122, 131, 179, 193, 230, 234, 260, 305

Building materials, 84-85, 91, 233, 236

Bunchberry, 260, 301

Bush pruning, 188-189

Business, 3, 9, 19, 239-240, 242-247

Butter, 149, 244, 288, 295

Butter bean, 152

Cabbage, 118, 120-121, 123, 125-133, 138-140, 152, 157-160, 205, 270, 285, 289

Calcium, 108, 190, 258, 266, 293

Canvas house, 85

Caraway, 122, 140

Career, 11, 38, 244

Carob, 260-261, 301

Carrot, 106, 110, 117, 120, 122-125, 127-132, 134, 137-139, 141, 149, 152, 157-158, 160, 170, 173, 186, 261, 286, 289, 305

Cash crops, 19, 22, 115, 173, 193, 217, 227-231, 235, 244-245, 248

Catfish, 69-71

Cattail, 101, 200, 261, 299, 301

Cattle, 23, 188, 200, 229, 266, 295, 296

    breeding, 215-216

    calving, 215

    forage, 231-233

    handling, 213-214

    milking, 216-217

    problems, 214

    scours, 207, 214

Cauliflower, 118, 132, 152, 158

Ceanothus, 261, 301

Cedar, 109, 126, 261-262, 305

Celery, 120-122, 124, 126, 128, 132, 134, 137, 140, 152, 173

Cellar, 35-36, 83, 141, 156-158, 161

Certification, 4

Chamomile, 119, 121, 123, 140, 173

Chard, 121, 123, 135, 140, 152, 160, 173, 193, 258

Charge controller, 47-49

Chayote, 152

Cherry, 185, 262, 294, 305

Chervil, 122-123, 126-127, 131, 134, 140

Chestnut, 172, 185, 232, 262, 298, 305, 308

Chick dust, 198

Chicken, 6, 23, 26, 175, 178, 187, 192-198

    breeds, 195

    chick dust, 198

    coop, 16, 36, 106, 194

    egg storage, 150

    eggs, 195-196, 230

    feed, 6

    forage, 36

    health, 194

    tractor, 193-195

Chicory, 109, 115, 158, 193, 200, 262, 305

Chinampa, 14-15, 68-69, 110

Chinese cabbage, 152

Chive, 123, 126, 131, 135, 140, 172-173, 284

Choosing plants, 115-134

Chutney, 165-166, 275

Cilantro, 123, 140, 172

Circuit breaker, 47

Circular system, 8

Cistern, 49, 62-63, 72

City garden, 135-136

Clamp, 157-158

Clarified fat, 149

Clay floor, 218-219

Gross head, 50-51

Ground storage, 158

Groundnut, 268, 302, 306

Guava, 185, 268, 306

Guilds, 115

Guyed tower, 53

Gypsum, 58, 67, 106, 108

Hairy vetch, 298, 302

Hawthorn, 235, 269, 302, 306

Hay, 46, 85, 96, 106, 110, 141, 174-177, 191, 194, 197, 199, 203, 204, 206, 210, 213, 216, 229-230, 233, 253, 254, 293

Hazelnut, 165, 269, 306

Head, 50, 51

Heart of palm, 285-286

Heat exchanger, 44, 46

Hedge, 20, 22, 24, 26, 33, 135, 170, 174, 179, 229, 232, 235, 243, 245, 259, 269, 271, 281, 294

Heel fly, 214

Hemp, 87, 126, 282,

Hemp insulation, 87

Herb business, 245

Hickory, 269, 306

High head, 50

High tunnel, 246

History, 3-5, 35, 191

Hive, 21, 179-184

Holm oak, 283

Holmgren, David, xiii, 3-5

Honey, 40, 148, 162, 164, 178, 180-183, 235, 244, 259

Honey locust, 193, 235, 269-170, 302

Honey super, 180

Honeysuckle, 270, 306

Hops, 135, 270-271, 306

Horseradish, 130, 161, 270-271, 306

Horseshoe dam, 64

Hot climate, 72, 78, 81, 145, 261

Hot water, 43-49, 82, 160

House, xii, 13, 16, 19, 20-26, 29-37, 43-61, 75-101
    appliances, 47, 82, 83, 145
    building materials, 84-92
    cooking, 145-148
    cooling, 33, 93

    design, 77-83
    earthbag, 84-85, 88-90
    earthquakes, 34-35, 85, 90
    energy-efficiency, 91
    flooding, 35, 83
    foyer, 77-78
    insulation, 42, 82, 85-87
    kitchen, 77-78, 81, 82, 90
    modification, 82-83
    passive solar, 9, 43-44, 49, 78, 82
    placement, 77, 242-243
    position, 36, 77
    protection, 33-35
    roof, 16, 33, 34-36, 47, 49, 53, 57-58, 60, 61
    snow, 32, 33, 36, 79
    temperature control, 43, 93-94
    underground, 88

Hugelkultur, 109-110

Human waste, 20, 99, 101

Humanure, 97-98, 101

Hurricane, 13, 24, 28, 35, 81, 172

Hydraulic pump, 50

Hydroelectric, 49, 51

Hyssop, 121, 126, 127, 131, 271, 306

Improving soil, 108-110

Income, xii, 9, 185, 231, 239-240

Industrial revolution, 5

Insulation, 33, 45, 82, 85-88, 92, 147, 148, 194

Intentional community, 239-242

Intercropping, 185-186

Inverter, 47-48, 53

Irish potato, 152

Irrigation, 10, 16, 19, 21, 22, 28, 57, 59, 60, 64, 66, 71, 72, 81, 170, 171, 187, 228, 263, 270, 296

Jack Rabbit wheel, 50

Jam, 163-166, 244, 245, 256, 258-261, 263, 266-267, 269, 270, 274-275, 278-279, 281, 285, 287, 289, 291-292, 294,

Japanese bath, 78-79

Japanese pagoda tree, 271-272

Jasmine, 272, 302, 306

Jerusalem artichoke, 152, 158, 174, 272-273, 306

Jicama, 273-306

culture, xii, 3-4, 240

definition, 3-4, 239

design process, 8-26

ethics, 5-8

for profit, 3, 9, 19, 239-240, 242-247

founders, 3-5, 242

history, 3-5

politics, 7, 239-240

principles, 5-8

religion, xiii, 240

Research Institute, 4

self-regulation, 37

success, 10-11

technology, 5, 7, 10, 240

theory, 3-8

Persian ground ivy, 303

Persimmon, 235, 289

Pest control, 191, 192, 245, 261, 264, 265, 278, 281, 287, 292, 305-308

Pests, 11, 39, 112, 115, 116, 121, 187, 190, 200, 252, 262, 264, 266, 267, 275, 279, 281, 282, 284, 289, 295

pH, 67, 69, 105, 108, 109

Photovoltaics, 47-49

Pickling, 150, 160

eggs, 150

Pig, 22, 23, 113, 116, 186, 187, 200-203, 220, 229, 273

breeding, 201-203

farrowing, 201

in permaculture, 200-201

Pigeon, 169, 170, 177-178

breeding, 177-178

in permaculture, 177

Pigeon pea, 193, 289, 303, 307

Pinto bean, 153

Pioneers, 12, 16, 105, 113, 227

Pizza oven, 148

Plant fiber insulation, 87

Plant guilds, 115

Plant indicators, 105, 108

Plants for cold, 138, 140

Plastic bags, 94, 150, 256

Plum, 164, 166, 185, 229, 281, 289, 294, 307, 308

Pole bean, 120, 121, 130

Politics, 7, 239-240

Pollution, 9, 24, 271

Polyculture, 15, 69, 71, 114, 228, 247

Polyethylene tank, 62-63

Pomegranate, 289-290, 303

Pond, 15, 19-20, 22, 25, 26, 34, 36-37, 59, 61-71, 88, 172-173, 182, 190-191, 198, 216, 261, 265, 290, 312

construction, 68

design, 65-66, 68

fertilizer, 14, 29,35, 69, 71, 263

fish, xi, 6-7, 12, 14-15, 19-20, 25, 61, 64-71, 96, 172, 190-191, 245, 265, 276, 311-312

management, 69

oxygen, 67-71

polyculture, 69, 71

population, 61, 66-67, 69-70, 198

waterproofing, 66

Poplar, 33, 232, 290, 303, 308

Potash, 108

Potato, 20, 96, 107, 110, 114, 118–120, 124, 125, 129–131, 133, 135, 138, 152, 156, 165, 173, 220, 290, 308

Poultry feed, 191

Power, 6-7, 9, 22-23, 36-37, 47-53, 57, 102, 153, 310-313

average usage, 48, 51, 53

battery, 47-48, 51, 53

calculating kWh, 51

DC, 47-50

solar, 23

water, 37, 49-51, 57, 153

wind, 51-54, 310

Prawns, 65, 69, 70,

Predators, 11, 68, 70, 116, 177, 191, 194, 195, 198, 212, 214

Preserving, 101, 141, 148-149, 156, 164, 242

butter, 146, 244, 288, 295

chutney, 165-166

cold storage, 83, 149-151, 156-157, 160, 269, 274, 276-277, 284, 286-287, 289-290, 311

drying, 77-78, 82, 150-155, 206, 208, 257, 260, 264, 269-270, 274, 276-279, 311 eggs, 18, 36, 39, 70, 115-116, 150, 177-178, 180, 183, 190-196, 198-200, 214, 230-231, 245, 312

fats, 149

freezing, 151

jam, 163-166

live, 156-158

meat, 148-149, 151, 153-155

safety, 148

salting, 149

sugar, 163-166

Pressure tank, 61

Prickly pear, 136, 290, 291, 303

Principles, xi-xii, 4-5, 8-9, 46, 65, 98, 106, 135, 221, 243, 245, 247, 310-311

diversity, 9,13,16, 38, 68, 112, 116, 235-236

functionality, xii, 10

native species, 10, 26, 261, 298

pollution, 9, 24, 271

work, 10

Products, 10, 17, 59, 88, 94, 151, 174, 239, 240, 244, 245, 247, 281, 283

Property map, 22, 24, 25, 184

Pruning, 188, 189

Pumpkin, 134, 130-131, 138, 151, 156, 234,

Purification, 61, 72-73

Pyramid pruning, 22, 188

Quail, 169-170, 177-178, 228, 245,

breeding, 177-178

in permaculture, 177

Queen bee, 181

Queen cell, 181-183

Quinoa, 193, 254, 291-292, 303, 308

Rabbit, 20,79,116, 126, 169-170, 174-177, 278

breeding, 175-177

in permaculture, 175

kindling, 176

Radish, 109-110, 117, 120, 122,124, 126-127, 129-132, 138, 158-161, 166, 292, 306

Rain, 24-25, 27, 28-29, 30, 35, 38, 57-60, 81, 95, 230, 261, 268

Rainwater, 20, 22, 25, 60-64

Rainwater catchment, 60-61

Raised bed, 106,108, 110, 137, 141, 228

Ram pump, 50

Rare plants, 245

Raspberry, 130,292, 303

Recycling, 8, 94, 248

Refrigeration, 101-102

Religion, xiii, 240

Replanting, 38, 138-139, 171, 200, 271, 273, 284, 292

Resources, 5, 7, 8, 9, 13, 17, 136, 239, 240, 247-249

Rhizobium, 39, 105

Rhubarb, 96, 166, 204, 292, 308

Rice, 40, 65, 67, 106, 165, 219-222, 254, 257, 263, 292, 295, 296, 299, 303, 304

paddy, 40, 219-221

Right livelihood, 243-244

Road, 19, 23-25, 34, 37, 58, 136, 169

Roof water collection, 60-61, 187,

Root vegetables, 110, 156, 157, 158, 213, 261, 265, 268, 272

Rose, 123, 126, 129, 131, 133, 134, 135, 164-166

Rosemary, 120, 121, 122, 131, 132, 134, 140, 166, 172-174, 179, 213, 292-293, 303, 308

Rotation 112, 141, 192, 205, 210, 220, 270, 295

Rubber floor mat, 219

Runoff, 25, 35, 57-5-, 63-65, 72, 137, 169, 187, 243,

Rush, 303

Russian olive, 193, 293, 303, 308

Rutabaga, 132, 153

Saddle dam, 63-64

Sage, 121, 122, 124, 131-134, 140, 172, 213, 293, 308

Salsify, 158, 293, 308

Salting meat, 149

Sand filter, 73-74

Sand floor, 219

Sandwich method, 90

Sauerkraut, 160, 161

Savory, 120, 132

Sawdust floor, 219

Sawdust insulation, 86

Sawdust toilet, 97-99

Schmutzdecke, 74

Scours, 207, 214

Scythe, 223-224, 230

Sector, 22-25, 30, 34, 36, 179

Spill gate, 59

Spillway, 64-65

Spinach, 106, 120-124, 125, 129, 131-134, 137, 153, 173, 257-258, 282-283, 303

Spiral, 11-14, 135, 169, 172-173

Squabs, 177-178

Squash, 115, 121, 124, 125, 128, 129, 130-133, 135, 151, 153, 156-157, 198, 221

Stage, 12, 13, 16, 105, 112-114, 222-223

Stanchion, 209

Staples, 20, 106, 137, 175, 249

Stone house, 85, 91-92

Stone pine, 295, 304

Storage tank, 44, 65, 73, 82, 170

Straw insulation, 86-87

Strawbale house, 85

Strawberry, 120-122, 127, 128, 132, 133, 295, 304

Stream braiding, 59

Sub-tropical climate, 28, 281

Success, 10-11

Succession, 12-13, 67, 101, 115, 137, 179, 192
    companion planting, 13

Sucker, 188

Sugar, 150, 155, 159, 161, 163-167, 179, 182, 184, 235, 260, 272, 273, 177, 178, 179, 293, 295

Sugar cane, 235

Sulfur, 108

Sumac, 295, 308

Summer squash, 153

Sun drying, 150, 153-155

Sunflower, 11, 124, 130, 133, 191-193, 213, 221, 234, 272, 295, 304

Sunlight, 18, 25, 29, 34, 82, 114, 170, 195, 270, 290

Sustainability, xi-xii, 5, 9-11

Swale, 22, 25, 34-36, 57-60, 65, 71, 72, 89, 108, 110, 137, 187, 194, 243

Sweet pepper, 153

Sweet potato, 151, 304

Sweet woodruff, 296

Sweetfern, 295-296, 304

Swept area, 51

Swiss chard, 135, 140, 160, 173, 193, 258

Tagasaste, 296, 304, 308

Tank, 15, 20, 25, 35, 44-47, 49, 57, 60-63, 65, 68, 70, 72-73, 79, 82, 97, 99, 101, 135, 170, 247

Tank aquaculture, 15, 61

Tansy, 121, 123, 126, 130-133, 186

Taro, 296, 304, 308

Tarragon, 125, 133, 140

Tea for soil, 110

Tea tree, 281, 308

Technology, 5, 7, 10, 240

Temperate climate, 262, 277, 288, 295

Temperature, 7, 17, 28-30, 33-34, 43, 46, 48, 57, 59, 65, 67, 69, 70, 83, 93, 95, 96, 99, 101, 12, 106, 111, 145, 146, 148, 149, 151, 156, 161, 181, 191, 195-197, 209, 242, 254-257, 261, 264, 266, 268-370, 275-277, 279-280, 287, 290, 294, 296, 297, 299

Thermal belt, 30

Thermal cooking, 146, 163

Thermal mass, 43

Thermosyphon, 44

Thresh, 221, 222, 225-226, 254, 260, 276, 288

Thyme, 121, 133, 134, 140, 172, 179, 197, 304

Tilapia, 68-71

Tilt-up tower, 53

Timber, 11, 232-233, 235, 253-256, 259, 261-262, 266, 269, 273-274, 277-278, 283, 285, 287, 289, 290, 293, 294, 298-299, 305-308

Tomato, 29, 34, 110, 117-119, 121-125, 127-130, 132-135, 137, 138, 140, 141, 154, 156, 157, 166, 170, 173, 198, 221, 247, 255, 261, 290, 292

Tornado, 6, 24, 35, 83

Tourism, 245

Tower, 53-54

Trees, 4, 11, 13, 15, 17, 19-22, 25, 30, 32-34, 38, 40, 53, 57-60, 65, 66, 81, 84, 89, 105, 106, 108-113, 116, 137, 169, 171-172, 184-188, 192-193, 200, 203-204, 210, 212, 229, 232-235, 243, 245, 248, 25-255, 265, 274, 276-278, 284, 287-289, 299

Trellis, 13, 14, 19, 20, 27, 32, 79, 80, 135-137, 170-171, 228, 257, 259, 267, 271, 274, 287, 288, 292

Trout, 69-71

Trust, 242

Turgo wheel, 50

Turnip, 128, 129, 132, 140, 141, 149, 153, 156, 158, 160, 173, 273, 297-298, 308

Underground house, 88, 92

U-pick, 228, 245,

Urban agriculture, 242, 247

Urban garden, 135-136

Urban sprawl, 248

Value added, 244-245

Vermicomposting, 96-97

Vetch, 113, 193, 221, 298, 302, 304, 308

Village, 239-240, 242-243

Village design, 242-243

Violet, 298, 403

Walnut, 109, 112, 126, 130, 134, 165, 185, 219, 235, 298, 308, 310

Washroom, 35, 78-79, 82, 90, 99

Waste, xii, 5, 9, 16, 20, 46, 53, 68, 73, 87, 89, 94, 95, 97-101, 109, 170, 192, 200, 235, 248

Water

    biomass heat, 43, 46-47

    calculating kWh, 51

    crops, 59, 65, 67

    dam, 16, 22, 25, 50, 59-67, 82

    desert, 56, 59-60, 65

    desert irrigation, 72

    distillation, 44, 48, 73, 235, 281

    diversion, 27, 71

    diversion drain, 59, 64, 65, 66, 71

    dry bed, 59-60, 206

    earthbag tank, 62

    elevation, 49-51, 63

    flood, 24, 25, 28, 35-36, 40, 58-60, 64, 65, 72, 82, 83, 100, 101, 222, 263, 264, 293

    floodplain, 24, 35, 60

    flow, 16, 26, 50, 58, 60, 67, 74, 108

    gravity fed, 10, 23, 62, 72, 82

    head, 50-51

    irrigation, 10, 16, 19, 21, 22, 28, 57, 59, 60, 64, 66, 71, 72, 81, 170, 171, 187, 228, 263, 270, 296

    lagoon, 60

    land area, 57-58

    multiple ponds, 67-68

    oxygen, 68-71

    parasite, 68, 73

    plants, 57-60, 65-72

    pond, 15, 19-20, 22, 25, 26, 34, 36-37, 59, 61-71, 88, 172-173, 182, 190-191, 198, 216, 261, 265, 290, 312

    pond management, 69

    power, 49-51, 57

    pressure tank, 61

    purification, 61, 72-75

    purifier maintenance, 73

    rain collection, 57, 60-62, 65, 71

    rainfall, 28-29, 57-59

    sand filter, 73-74

    small pool, 60

    soil impact, 38

    solar heating, 43-49

    sources, 49-50

    storage tank, 44, 65, 73, 82, 170

    stream braiding, 59

    swale, 22, 25, 34-36, 57-60, 65, 71, 72, 89, 108, 110, 137, 187, 194, 243

    tank algae, 61, 69-70, 74

    viability, 51

    waste, 64-65

    waterproofing, 47, 59, 62-66, 91, 148, 157, 284,

Water chestnut, 65, 67, 172, 298-299, 308

Watercress, 172, 299, 304

Wax bean, 153

Weeds, 16, 33, 39, 68, 96, 106, 107, 112, 113, 115, 174, 190, 193, 198, 200, 221-223, 260, 275, 296

Wild ginger, 299,304

Wild rice, 106, 299, 304

Wilderness, 11, 19, 22, 111, 236, 242,

Wildlife corridor, 11, 22, 236

Willow, 34, 68, 179, 232, 235, 245, 299-300, 304, 308

Wind, 11, 16-20, 22-24, 27-28, 30-38, 48-55,105, 111, 116, 169, 172, 174, 179, 184, 191, 217, 242, 247, 256, 285

    calculating kWh, 51

    hurricane, 13,24,28,35, 81, 172

    power, 51-54

    tower, 53-54

    turbine installation, 53-54